Collins

Revision

NEW GCSE

Chemistry

OCR

Twenty First Century Science

Authors: Brian Cowie
Ann Tiernan

Revision Guide +
Exam Practice Workbook

Contents

The changing air around us

The gases that make up the air

- Air (or **atmosphere**) is a **mixture** of **gases**:
  - it is mainly nitrogen, oxygen and argon
  - it contains small amounts of water vapour (H_2O) and **carbon dioxide** (CO_2).
- Clouds are water or ice and dust is a solid, so they are not parts of the air.

- Gases spread out to take up all the space available.
- Particles are very small, so in gases there is lots of empty space between gas **molecules**. This means gases can be squeezed into a smaller volume, like in a bike tyre.
- Dry air contains: 78% nitrogen (N_2); 21% oxygen (O_2). The remaining 1% is mainly argon (Ar) and very small amounts of other gases.

Most of a gas is empty space.

EXAM TIP
When drawing diagrams to show particles in air, remember that most of the available space will be empty.

Remember!
Two gases make up 99% of the air. All the rest fit into the last 1%.

- Oxygen gas reacts with most metals to make solid metal oxides.
- We can find the percentage of oxygen in air by passing air over heated copper.

The Earth's atmosphere

- Earth's atmosphere was probably formed about 4 billion years ago by gases given out by volcanoes.
- Volcanoes release huge amounts of carbon dioxide and water vapour. They also release **lava** and dust.

- Different processes have removed almost all of the carbon dioxide that was in the early atmosphere, leaving the air with the composition we have today:
  - Four billion years ago the Earth's atmosphere was very hot.
  - As the Earth cooled, oceans formed from the condensed water.
  - About 3 billion years ago simple **bacteria**-like creatures evolved to use **photosynthesis**.
  - This removed carbon dioxide from the air, and released oxygen, allowing animals to evolve.
  - Carbon dioxide was removed by plants and animals dying and becoming buried.
  - Over millions of years some of the buried material became **fossil fuels**.
  - Carbon dioxide **dissolved** in oceans reacts with salts to form insoluble calcium carbonate.
  - This forms **sediments** which become buried and cemented to form **sedimentary** rocks.

- Ideas about the composition of the Earth's atmosphere have changed over time:
  - Sixty years ago many scientists thought the early atmosphere was largely ammonia and **methane**.
  - Recent rock composition discoveries showed early ideas were not correct, and the early atmosphere was largely carbon dioxide.

Ideas about science

You should be able to:
- suggest reasons why collecting air quality data at the same location may give different values
- understand why taking several measurements of different air pollutants gives the best estimate of their true value.

Improve your grade

The Earth's atmosphere
Foundation: The early atmosphere was mainly composed of water and carbon dioxide.
Suggest how these were gradually removed.

AO1 [4 marks]

C1 Air quality 3

Humans, air quality and health

How has human activity changed air quality?

- Humans are changing the gases in the atmosphere by burning fuels. Fuels are used in factories, power stations, for transport and in homes.

- Gases we call **pollutants** are harmful to health. Examples are **carbon monoxide**, **nitrogen oxides** and **sulfur dioxide**.

- Pollutants are harmful to the environment and to the people and animals living there. For example, carbon monoxide reduces the amount of oxygen blood can carry.

- Air quality is 'good' if it has very few pollutants and 'poor' if there are lots of pollutants.

- Burning fuels releases carbon dioxide and solid **particulates** that float in the air, e.g. **carbon** (soot). Particulates are also released naturally as ash from volcanoes.

- In the last 50 years the amount of carbon dioxide in the air has increased by 25%.

- Carbon dioxide is linked to climate change, by acting as a 'greenhouse' trapping heat in the atmosphere.

- Human activity, like burning down forests to make more farmland, increases carbon dioxide and particulates.

- When air pollution levels are high, more deaths from asthma, heart disease and lung disease occur. There is a **correlation** between air quality and health.

- Air quality is more of a problem in large cities, such as Mexico City and Beijing. Some countries, e.g. the UK, have made laws to try to improve air quality.

- Sulfur dioxide and nitrogen dioxide are pollutants that make **acid rain**, which damages plants and animals.

- Sulfur dioxide and nitrogen oxides are asthma triggers for some people.

> ### EXAM TIP
> Make sure you know the names of the two gases responsible for acid rain. Do not get this mixed up with carbon dioxide, which is linked to global warming.

Measuring air quality

- Small amounts of carbon dioxide in the air are measured in parts per million (ppm). 1 ppm means that there is 1 gram of the pollutant substance in 1 million grams of air.

- The other pollutant gases – carbon monoxide, nitrogen oxides and sulfur dioxide – are measured in parts per billion (ppb).

- The amount of pollutant gases are measured in air quality monitoring stations throughout the UK. The data is transmitted automatically to a central computer for analysis.

> **Remember!**
> Some pollutants *directly* affect us, but some like acid rain harm the environment, so harm us *indirectly*.

Correlation and cause

- A correlation is a link between a factor and an outcome. Data is needed to show this.

- To establish a causal link, evidence needs to show that changing a particular factor is the only cause of a particular outcome. For example, when the levels of sulfur dioxide increase, statistics show that more people have asthma attacks. The data shows a correlation or link, but sulfur dioxide is not the only asthma trigger. Some asthmatics have attacks for completely different reasons.

Ideas about science

You should be able to:

- in a given context, identify the outcome, e.g. asthma cases, and the factors that may affect it, e.g. increase in air pollution

- in a given context, suggest how an outcome, e.g. car emissions, might alter when a factor is changed, e.g. stricter MOT testing

- identify, and suggest from everyday experience, examples of correlations between a factor and an outcome where the factor is (or is not) a plausible cause of the outcome.

Improve your grade

How has human activity changed air quality?
Foundation: How do humans affect air quality?

AO1 [4 marks]

Burning fuels

What happens when fuels burn?

- Oxygen is needed for any fuel to burn and release energy.
- Fossil fuels such as petrol, diesel and fuel oil are mainly **hydrocarbons**. Hydrocarbons only contain carbon and hydrogen atoms.
- Coal is a fossil fuel mainly made of carbon atoms.

- When a hydrocarbon fuel burns: hydrocarbon fuel + oxygen → carbon dioxide + water (+ energy)
- **Oxidation** is when oxygen is added to a substance.
- **Reduction** is when oxygen is removed from a substance.
- **Combustion** (or burning) is an oxidation reaction.

EXAM TIP

If a question asks for a word equation, do not try to use symbols. Some words to use will probably be in the question.

- Gases in the atmosphere can be separated.
- Pure oxygen makes fuels burn more rapidly and at higher temperatures. One example is in an oxy-fuel welding torch, which can melt steel.

What happens to atoms in chemical reactions?

- **Atoms** do not change. In chemical reactions atoms get rearranged to make new substances.
- Atoms of non-metal elements join to form **molecules**.

- **Elements** are rearranged to make new **compounds** in chemical reactions.
- Atoms in **reactants** are rearranged into new **products** with different properties.

Remember!
The numbers and types of elements in the reactants are the same as in the products.

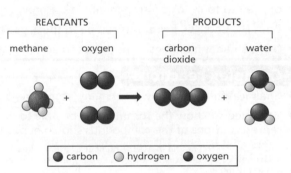

REACTANTS — methane   oxygen   PRODUCTS — carbon dioxide   water

● carbon   ○ hydrogen   ● oxygen

Methane burns in air to form carbon dioxide and water.

- Mass is conserved in a reaction because all the atoms in the reactants are just rearranged into the products. For example, all the carbon atoms in all the fossil fuels ever burnt are still present, but in different forms.

Burning sulfur

- Solid yellow sulfur burns and makes a colourless gas called sulfur dioxide.
- Sulfur is insoluble but sulfur dioxide dissolves in water make an acid solution.

- Fossils fuels contain small amounts of sulfur from the plants and animals that formed them.
- When fossil fuels burn the sulfur burns: sulfur + oxygen → sulfur dioxide ($S + O_2 → SO_2$)
- Coal often contains the most sulfur, so burning coal can give off more sulfur dioxide than other fossil fuels.

- In the 1970s people noticed forests and aquatic life in ponds were dying.
- Scientists were able to use data to explain that sulfur dioxide caused acid rain.
- Acid rain lowers the pH when it falls on land or enclosed water, harming living things or eroding carbonate rock.
- Acid rain does not affect humans directly, so it is called an indirect pollutant.

◉ Improve your grade

What happens when fuels burn?

Foundation: Methane (CH_4) reacts with oxygen to form carbon dioxide and water. Finish the diagram to show this reaction. Use ● to represent a carbon atom, ● to represent a hydrogen atom and ○ to represent an oxygen atom.

methane   +   oxygen   →   carbon dioxide   +   water

AO2 [3 marks]

Pollution

How pollutants are formed

- Power stations and transport make most pollution because they burn most fuel. Electricity production and transport has increased over the last century.
- Sulfur dioxide is made if the fuel contains sulfur.
- Carbon dioxide is always formed when fuels burns.
- If not enough air is available to burn the fuel:
  - poisonous carbon monoxide is made
  - bits of solid carbon (soot) called particulates are made, making surfaces they land on dirty.
- Car engines make nitrogen oxides when nitrogen and oxygen from the air react at high temperatures. This contributes to acid rain.
- Air quality measurements need repeating many times because results vary. On dry, hot, calm days, air pollutants can be trapped in cities.

- Carbon monoxide and particulate carbon are formed during incomplete combustion (insufficient oxygen).
- Only in the last 50 years have scientists discovered how different air pollutants form, and how they react with air to produce **smog**, acid rain and climate change.

Higher
- Nitrogen monoxide (NO) is formed in furnaces and engines at a temperature of about 1000 °C.
- When nitrogen monoxide is released into the atmosphere it cools. It then reacts with more oxygen to form toxic nitrogen dioxide, a brown gas.
- Both NO and NO_2 pollutants can be in the air, so NO_x is used to represent both of them.
- NO_x damages buildings, contributes to acid rain, and can affect health.

> **Remember!**
> Fossil fuels all contain impurities, which when burnt can produce pollutants.

> **EXAM TIP**
> Make sure you know the different between *complete* and *incomplete* combustion, and know the products formed.

Complex reactions

- Chemical formulae show how many atoms are joined together.
- You need to know the formula and be able to draw visual representations of the compounds shown opposite.

The formula of a compound shows the number of atoms of each element that are joined together in a molecule. The molecule diagrams show the arrangement of the atoms in the molecule.

CO_2 CO H_2O SO_2 NO NO_2

● carbon ○ hydrogen ● oxygen ○ sulfur ● nitrogen

Removing pollutants and monitoring air quality

- Atmospheric pollutants don't simply disappear – they have to go somewhere.
- Pollutants are removed from the air when:
  - particulate carbon settles on surfaces, making them dirty
  - sulfur and nitrogen oxides react with water and oxygen to produce a mixture of sulfuric acid and nitric acid in rain (acid rain)
  - carbon dioxide is used by plants for photosynthesis
  - carbon dioxide dissolves in rain water and in oceans.
- Climate scientists take the **mean** of many measurements for each pollutant. The mean is a good estimate of true value.
- The **range** is the difference between high and low results.

Higher
- Air quality results vary for several reasons. Potential **outliers** could actually be valid data, and removing them could lead to mistakes.

Ideas about science

You should be able to:
- from a set of repeated measurements of a quantity, use the mean as the best estimate of the true value
- explain why repeating measurements leads to a better estimate of the quantity
- from a set of repeated measurements of a quantity, make a sensible suggestion about the range within which the true value probably lies and explain this.

Improve your grade

How pollutants are formed

Foundation: Describe the damage acid rain causes, and explain why the UK is being blamed for acid rain damage in Europe.

AO1 [4 marks]

Improving power stations and transport

Improving power stations

- Reducing electricity reduces fossil fuel use in power stations.
- New electrical products use less electricity, but some is wasted if they are left on standby.

- Burning oil and gas makes less sulfur dioxide than burning coal.
- Sulfur can be removed from oil and gas before it is burnt, but it is harder to remove from coal.
- Power stations are developing ways of reducing pollution by cleaning waste gases.
- Power stations can remove solid particulates using electrostatic filters.
- Sulfur dioxide can be removed from waste gases by **flue gas desulfurisation**.

- Two 'wet scrubbing' methods used to remove sulfur dioxide from power station waste gases are:
  - **1** using an alkaline slurry of calcium oxide (lime) and water to make gypsum (calcium sulfate), which can be sold as plaster
  - **2** using sea water, a natural alkaline which absorbs sulfur dioxide.

Higher

Reducing CO_2 and replacing fossil fuels

- Burning less fossil fuel reduces the amount of carbon dioxide (CO_2) gas released.
- Ways to reduce our use of fossil fuels include: using alternative energy sources; improving building insulation; walking, cycling, using public transport.

- One alternative to fossil fuels is **biofuels**, which are made from plants. Examples are wood chips, palm oil and alcohol made from sugar.
- Biofuels are 'carbon neutral' – when they are burned they release the same amount of carbon dioxide that the plant originally took from the air to grow.
- Large areas of land are needed to grow biofuels. The land could be used for growing food.
- Gas produces less carbon dioxide than coal for the same amount of energy released.
- Fossil fuels are not renewable, so they are not a **sustainable** source of energy.

Remember!
To meet the same energy demand, we need to burn less fossil fuels and find alternatives. This will also reduce pollution.

Reducing air pollution from transport

- Air pollution from vehicles can be reduced by:
  - using cars less, especially for short journeys
  - using cleaner fuels and removing pollutants from exhausts
  - making public transport cheaper, more frequent and available in more places.

- Modern vehicles have more efficient engines that use less fuel.
- **Catalytic convertors** contain a platinum catalyst that allows pollutant gases to react with each other: carbon monoxide + nitrogen monoxide $\longrightarrow$ nitrogen + carbon dioxide. In this reaction:
  - carbon monoxide gains oxygen so it is **oxidised**
  - nitrogen monoxide loses oxygen so it is **reduced**.
- Low sulfur fuels are needed as sulfur damages the catalyst. Using low sulfur fuels also reduces sulfur dioxide emissions.
- Legal limits for exhaust emissions are enforced by strict MOT tests.

- Electric cars do not give out pollutant gases when being used, but the electricity produced by fossil fuel power stations used for charging does.
- Research continues into improving batteries and improving charging times. Few charging points are available at present.

EXAM TIP
In questions that start with 'Suggest', try to include a wide range of answers.

Higher

Improve your grade

Reducing air pollution from transport
Foundation: Suggest ways of reducing air pollution from cars.

AO1 [4 marks]

C1 Summary

99% of the Earth's atmosphere is mainly nitrogen and oxygen. The remaining 1% is mainly argon, with tiny amounts of carbon dioxide and other gases.

The Earth's atmosphere

Pollutants are released by natural processes such as volcanoes and by human activity such as burning fuels.

Earth's atmosphere was probably formed by volcanic activity.

Carbon dioxide dissolved in the sea and became trapped in sedimentary rocks and fossil fuels.

Changes in the atmosphere

As the Earth's first atmosphere cooled, water vapour condensed to form oceans.

Plants evolved. Photosynthesis removed carbon dioxide and released oxygen.

Elements are made of atoms. Non-metal atoms combine to make molecules.

Air pollution

UK air pollution levels are regularly monitored.

Chemical reactions involve arranging atoms into new substances which have different properties.

Coal is carbon.

In a plentiful air supply it burns to form carbon dioxide.

In limited air it burns to form poisonous carbon monoxide. Some do not burn and makes soot.

Carbon monoxide, sulfur dioxide, nitrogen dioxides and particulate carbon (soot) are all air pollutants.

Nitrogen oxides are made when nitrogen and oxygen from the air react in hot engines and furnaces.

Making air pollution

Small amounts of sulfur are in all fossil fuels.

When the fuel burns, the sulfur forms sulfur dioxide.

Sulfur dioxide dissolves in water forming acid rain.

Power stations and transport make the most pollution as they burn most fuels.

New electrical products are designed to save energy, reducing electricity demand.

We can reduce pollution by not using cars as much, using cleaner fuels, and having cars serviced regularly.

Catalytic convertors remove nitrogen oxides and carbon monoxide from exhaust fumes.

Carbon monoxide is oxidised into carbon dioxide.

Nitrogen oxides are reduced into nitrogen.

Improving power stations and transport

Power stations can reduce pollution by flue gas desulfurisation.

Biofuels are made from plants and are classed as 'carbon neutral'. They can be used in cars and by power stations.

Using and choosing materials

Comparing materials and measuring properties

- Each material has properties that make it suitable for the job it is doing.
  - Rubber is used for car tyres because it hard and **elastic**.
  - **Fibres** are used to weave cloth into clothes.
  - **Plastics** keep their shape when moulded into objects like washing-up bowls.
- Different plastics have different properties. Manufacturers make products by choosing plastics that give the best properties.

- Properties describe how a material behaves.
  - **Melting point** is the temperature at which a solid turns into a liquid.
  - **Tensile strength** is the force needed to break a material when it is being stretched.
  - **Compressive strength** is the force needed to crush a material when it is being squeezed.
  - **Stiffness** is the force needed to bend a material.
  - **Hardness** is how well a material stands up to wear. Hardness can be a compared by scratching two materials together.
  - **Density** is the mass of a given volume of the material. It compares how heavy something is for its size.

- Some properties depend on the size and shape of the material being tested.
- Density is mass per unit volume. The units are g/cm^3 or kg/m^3.
- The effectiveness and durability of a product depend on the materials used to make it:
  - Some materials can be drawn into thin filaments with greater tensile strength. They can be spun into fibres and woven into cloth.
  - Ropes are made by winding fibres together. The more that are wound, the greater the strength.
- Rubber is an elastic material that bounces back when a force is removed. Different types of rubber have different compressive strengths and hardnesses.

EXAM TIP
Questions about properties often give data in tables. Take time to understand what the data is showing before answering the questions.

Errors and variation in measurement

- A single result may vary, so repeats are needed.
- A result which is very different might be an **outlier** – an incorrect result.
- Calculating the **mean** (average) is a good way to estimate the **true value**.

- Many measurements need to be taken to find the true value.
- The **range** is the smallest to the largest result, excluding outliers.
- We can never be sure if a set of measurements gives the true value.

- **Errors** in measurement produce variations in data.
- Outliers can only be discarded if an error occurred in the measurement.

Higher

Ideas about science

You should be able to:
- suggest reasons why several measurements of the same quality may give different values
- when asked to evaluate data, make reference to its repeatability and/or **reproducibility**
- estimate the true value.

Improve your grade

Comparing materials and measuring properties
Foundation: Sub aqua divers should always leave a marker buoy at the dive site to warn other boats to stay away, reducing the risk of divers being hit when they surface.
Describe the main properties that the marker buoy would need.

AO1 [4 marks]

C2 Material choices   **9**

Natural and synthetic materials

Natural and synthetic materials

- All the materials we use are chemicals or mixtures of chemicals.
  - **Metals** are chemicals which are shiny, **malleable** and electrical conductors.
  - **Ceramics** include clay, glass and cement. They are hard and strong.
  - **Polymers** are large molecules used to make rubbers, plastics and fibres.
  - Concrete is a mixture of sand and cement.
  - Bronze is a mixture of copper and tin.

A nylon fibre is synthesised by reacting chemicals in two solutions together. The solutions do not mix and the nylon is formed at the interface between them.

- **Natural materials** from living things which need little processing are cotton and paper from plants and silk and wool from animals.

- Other natural raw materials which are extracted from the Earth's crust are limestone, iron ore and **crude oil**.

- **Synthetic materials** are manufactured by chemical reactions using raw materials.

- Synthetic materials are alternatives to natural materials from living things.

- Synthetic materials have replaced natural materials because:
  - some natural materials are in short supply
  - they can be designed to give particular properties
  - they are often cheaper and can be made in the quantity needed.

> **Remember!**
> Synthetic materials are made by chemical reactions, and can be designed to do particular jobs.

Crude oil and using hydrocarbons

- Crude oil (petroleum) is a mixture of thousands of **hydrocarbons**. Hydrocarbons are compounds of just carbon and hydrogen atoms.
  - Most hydrocarbons from crude oil are used as fuels.
  - When fuels burn in oxygen, carbon dioxide and water are made (see diagrams below and on page 5).

(see diagrams below and on page 5)

- Burning a fuel like methane is a chemical reaction, so atoms are rearranged into new products.

- The number of atoms of each element in the reactants must be same in the products.

> **EXAM TIP**
>
> When representing a chemical reaction, count the atoms to make sure the reactants and products contain the same number and same types.

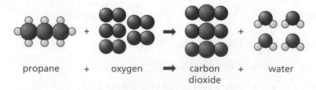

| propane | + | oxygen | ⇒ | carbon dioxide | + | water |

Propane burns in air to produce carbon dioxide and water.

- Crude oil consists mainly of a mixture of hydrocarbons, which are chain molecules of varying length up to 100 carbon atoms long.

- As crude oil is a mixture, its composition varies from place to place.

name	methane	ethane	propane
formula	CH_4	C_2H_6	C_3H_8
atomic model			

Some of the alkanes.

- Nearly 90% of crude oil is used as fuels.

- Around 3% of crude oil, mainly smaller hydrocarbon molecules, is used to synthesise other chemicals. Examples of synthesised chemicals from oil are ethanol and plastics.

Improve your grade

Natural and synthetic materials

Higher: What are synthetic materials, and what advantages do they have over natural materials?

AO1 [4 marks]

Separating and using crude oil

Separating out the substances in crude oil

- Crude oil is separated by **fractional distillation**:
  - The oil is heated up which turns it all into gases.
  - The distillation tower gets cooler as it gets higher.
  - Gas molecules **condense** into liquids when they cool.
  - Liquids with similar boiling points collect together. We call these **fractions**.

- Hydrocarbons in each fraction have boiling points within a range of temperatures.

- Molecule chain lengths are similar sizes within each fraction.

- The smaller the molecule chain length, the lower the boiling point.

- The smaller the molecule chain length, the smaller the forces between molecules.

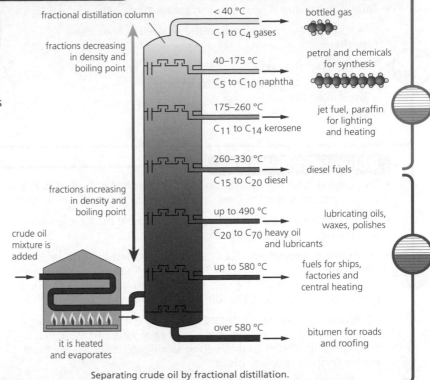

Separating crude oil by fractional distillation.

Investigating boiling points

- Attractive forces exist between molecules in crude oil, holding them together.

- As the hydrocarbon chain length increases, the force between these molecules increases.

- Larger molecules need more energy to break them out of a liquid to form a gas, so have higher boiling points.

Remember!
The lower the boiling point, the smaller the molecule chain length, and the higher it will rise during fractional distillation.

Making polymers

- A **polymer** is a large molecule made by joining many smaller molecules called **monomers**. A polymer can have a chain of anything from hundreds to millions of carbon atoms.

- A polymer is made by a process called **polymerisation**.

- Polymers with better properties mean some older materials have been replaced. Examples are plastic buckets and carbon fibre tennis rackets.

- Ethene is a monomer used to make polyethene.

- Different monomers produce different polymers.

- PET (polyethylene terephthalate) is a polymer used to make drinks bottles. PET is clear, strong, has a low density and does not shatter. This makes it a superior material to glass.

- Polymer chains can be altered by replacing hydrogen atoms with other atoms or groups of atoms.

- Each new polymer has its own set of properties and uses.

- Materials such as Kevlar have advantages over alternatives, but can also have disadvantages.

- Material choice will depend on comparing properties for different jobs, with cost being a factor.

Improve your grade

Investigating boiling points

Higher: Explain why during the distillation of crude oil, small hydrocarbon molecules rise to the top of the tower.
AO1 [4 marks]

Polymers: properties and improvements

Attractive molecules

- Small forces attract molecules to each other.
- The forces are strongest when the molecules are close together.
- The stronger the force:
  - the more energy is needed to separate the molecules
  - the higher the melting point.

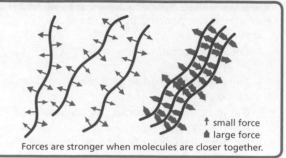

↑ small force
⬆ large force

Forces are stronger when molecules are closer together.

Polymer differences

- Polymers can be made with properties that make them suitable for a range of different uses.
- The properties of polymers depend upon how their molecules are arranged and held together.
- Low density polyethene (LDPE) has long molecules with branches. The branches keep molecule chains apart, so the forces between different molecules are weak. Items made from LDPE, e.g. plastic carrier bags, are weak, flexible, soft and have low melting points.
- High density polyethene (HDPE) has long chains but no branches, so the molecules are aligned close to each other. HDPE is much stronger and is used to make long-lasting items which are hard and stiff, such as water pipes.

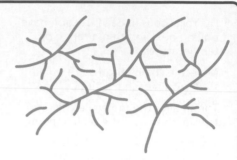

LDPE has branches between the molecular chains, which reduce the attractive forces between them.

- **Higher** HDPE has a high degree of crystallinity. This means there are lots of areas with regular patterns in the way the molecules line up.
- High **crystalline** polymers are strong with high melting points, but can be brittle.

Improving polymers

- Making the molecule chain longer makes it stronger.
- Longer chains need more force to separate them.
- Longer chains have higher melting points than short chains.

- **Plasticisers** are used to make a polymer softer. They are small molecules inserted into polymer chains to keep them apart, weakening the forces between them.
- Plasticised **PVC** is still hardwearing and waterproof, but it is also flexible, making it a suitable material for rain coats.
- **Thermoplastics** soften when heated and can be moulded into shape.
- **Thermosetting** plastics do not soften when heated. They contain **cross-links** which lock the molecules together so they cannot melt.

- Crystallinity can be increased by removing branches on the main polymer chain and making the chains as flat as possible. This is so that the molecule chains can line up neatly.
- **Higher** Drawing polymers through a tiny hole when heated makes the molecule chains line up, increasing crystallinity and forming a higher tensile strength fibre.
- Materials which have been treated in this way include bullet-proof vests and sail material (Kevlar).

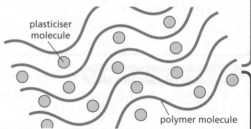

plasticiser molecule

polymer molecule

The plasticiser molecules make the material less rigid.

Remember!

All plastics can be classified as either thermosetting (remains rigid when set) or thermoplastics (melt easily when heated).

Improve your grade

Improving polymers

Higher: A company is making rotor blades for a radio-controlled toy helicopter.

A polymer needs to be made stronger but more flexible. What could be done, and how will it change the properties?

AO2 [3 marks]

Nanotechnology and nanoparticles

Small natural nanoparticles

- The width of a human hair is about 0.1 millimetres.
- Microscopes are used to view small objects like human cells.
- Molecules and atoms are thousands of times smaller still.
- **Nanoparticles** are materials containing up to a thousand atoms.
- Nanoparticles:
  - occur naturally, such as salt in seaspray
  - occur by accident, such as solid particulates made when fuels burn
  - can be designed in laboratories.

Nanotechnology

- **Nanotechnology** is the use and control of very small structures. The size of these structures is measured in **nanometres (nm)**. A nanometre is one millionth of a millimetre.
- An atom is about one-tenth of a nanometre in diameter.
- Nanoparticles can be built up from individual atoms. These structures are about the same size as some molecules.
- 'Buckyballs' (see opposite) are very strong carbon spheres made of 60 carbon atoms.
- Carbon **nanotubes** are being designed in laboratories.
- Some nanoparticles are effective **catalysts** as they have a large surface area. Increasing surface area provides more sites for reactions to take place.
- Surface area increases when a lump of solid is cut up into bits.

A model of a natural nanoparticle of carbon called a 'buckyball'.

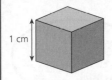

volume 1 cm³
surface area 6 cm²

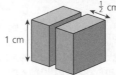

volume 1 cm³
surface area 8 cm²

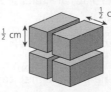

volume 1 cm³
surface area 10 cm²

1 cm³ of gold could make 1×10^{21} nanoparticles, each with a volume of 1 nm³. If each nanoparticle was a cube it would have a total surface area of 6×10^{21} nm², which is 6×10^{6} cm². This is one million times the surface area of the original piece of gold.

- Nanotechnology builds structures from 10 atoms across (1 nm) up to a thousand atoms across (100 nm).
- While the diameter of nanotubes is measured in nanometres, they can be millimetres long.
- Nanoparticles have very large surface areas. Because of this, they show different properties to larger particles of the same material.

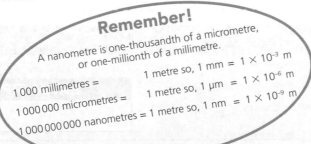

Remember!

A nanometre is one-thousandth of a micrometre, or one-millionth of a millimetre.

1 000 millimetres = 1 metre so, 1 mm = 1×10^{-3} m

1 000 000 micrometres = 1 metre so, 1 μm = 1×10^{-6} m

1 000 000 000 nanometres = 1 metre so, 1 nm = 1×10^{-9} m

These nanostructures made from carbon atoms are called nanotubes. They may be as small as 3 nm in diameter.

Improve your grade

Nanotechnology

Foundation: Explain what nanoparticles are, and suggest why some act as catalysts.   *AO1* [4 marks]

The use and safety of nanoparticles

Using nanoparticles

- Silver nanoparticles are very good at killing bacteria. They can be: added to fibres and woven into socks; put into wound dressings; put into plastic and made into food containers.

- Titanium oxide nanoparticles are put into sunscreen. They make the sunscreen **transparent** (no white residue) and absorb UV light.

- Nanoparticles can be mixed with other materials like metals, ceramics and plastics. These combined materials are called **composites**.

- Composite materials are stronger and harder wearing. Adding nanoparticles to:
  - plastic sports equipment makes it stronger
  - tennis balls make them stay bouncy for longer
  - rubber used in tyres make them harder wearing.

- Nanotechnology is the science of making and using nanoparticles.

- Graphite forms in strong sheets that separate easily. Individual graphite sheets one-atom thick are known as graphene sheets.

- Graphene sheets can be rolled into carbon nanotubes. These are super-strength materials.

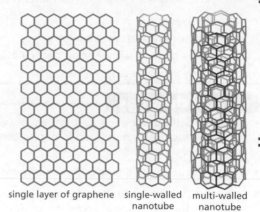

single layer of graphene   single-walled nanotube   multi-walled nanotube

Modern composite technology using graphene nanotubes results in super-strength materials.

Are nanoparticles safe?

- Silver nanoparticles can be washed out of clothes containing them and get into sewage works.

- Sewage works use bacteria to clean water. Silver nanoparticles could kill these useful bacteria.

- If silver nanoparticles are released into the environment they could kill lots of useful microorganisms.

- Nanoparticles are also used in cosmetics and sunscreens. The nanoparticles are added to materials that have already been used and tested.

- Nanoparticles are small enough to pass through skin into blood, and into body organs. The possible medical effects of this are not yet known.

- While a lot of research is taking place into the use of nanoparticles, little has been carried out into their possible harmful effects.

- One fear is that nanoparticles in the air might be breathed in and cause lung or brain damage.

Remember!
Nanoparticles may have risks as well as benefits and both need to be considered.

- Some people think that because nanoparticles occur naturally, such as in soot and volcanic dust, they pose no danger.

- Others disagree because new nanoparticles with new properties have been manufactured.

- No one knows if nanoparticles used in solids, like windows and paintwork, can escape into the air.

- Some people want proof that new nanotechnologies will not create health and environmental **risks**.

- Risk is defined as the change of an event occurring, and the consequences if it did.

Ideas about science

You should be able to:

- explain why it is impossible for anything to be completely safe

- identify examples of risks which arise from nanotechnology

- suggest ways of reducing a given risk.

EXAM TIP

If asked to identify risks and benefits, use different examples for each one.

Improve your grade

Are nanoparticles safe?

Higher: Fresh Crop is a company selling mixed salads. They are considering adding silver nanoparticles to their food packaging to help prevent bacterial decay.

Explain why some people believe this may have risks.   *AO2* [3 marks]

C2 Summary

Measuring the properties of materials

Materials like plastics, fibres and rubber have many different properties. All these need to be considered when choosing them for a job.

Some properties are melting point, strength, stiffness, hardness and density.

Materials vary slightly, so they need to be measured many times to establish the range, eliminate outliers, find the mean, and estimate the true value.

Natural and synthetic materials

Synthetic polymers are man-made materials from the Earth's crust. Examples are:
• plastics from crude oil
• aluminium metal from bauxite ore.

Natural polymers come from living things. Examples are:
• cotton from plants
• silk from animals
• limestone and iron ore from the Earth's crust.

Synthetic materials are alternatives to using natural materials. They include:
• plastic washing-up bowls instead of metal or ceramic bowls
• neoprene wetsuits instead of rubber ones
• Kevlar body armour instead of metal plates.

Separating crude oil

Crude oil is made of hydrocarbon chains of varying lengths.

Crude oil is separated by fractional distillation into fractions.

Larger hydrocarbons have more attractive forces between the molecules, so more energy is needed to separate them, resulting in higher boiling points.

Different fractions are used mainly as fuels and lubricants, with a small amount used for raw materials, to make materials like plastics.

Each fraction contains a small range of molecule sizes of similar boiling points.

Plastics

The strength, stiffness and hardness of a plastic links to the amount of energy needed to separate the molecule chains.

Small molecules called monomers can join together to make polymers.

Polymer strength depends on chain length, cross-linking, plasticisers and increased crystallinity.

Nanotechnology

Nanotechnology involves very small structures up to 100 nm in size.
They occur naturally, by accident, and by design.

Nanoparticles have different properties compared to larger particles of the same material. One key factor is their larger surface area compared to their volume.

Carbon nanoparticles can be used to strengthen sports equipment and body armour.
Silver nanoparticles give fibres antibacterial properties.

The health effects of using nanoparticles are unknown as they have not been around very long.

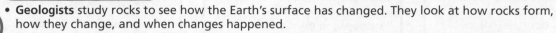

Moving continents and useful rocks

Looking at rocks

- **Geologists** study rocks to see how the Earth's surface has changed. They look at how rocks form, how they change, and when changes happened.
- Geological changes happen by slow movements of **tectonic plates**. Plates can move by sliding past each other, colliding or pulling apart.

- Plate collisions build mountain ranges, which erode over time.
- Geologists can explain most of the past history of the Earth by processes they can observe today.

Making Britain

- Over millions of years, Britain has moved across the Earth's surface.

- 600 million years ago, England and Wales were separated from Scotland by ocean, and both were near the South Pole.
- Gradually, different continents drifted and crashed together to form a **supercontinent**, Pangea.
- Britain is made from rocks from different ancient continents.
- Originally, Britain was nearer the equator with a warmer climate.
- Different climates existed in Britain, from tropical swamps to ice ages.

Remember!
The UK has many different rocks and raw materials due to continental movement and climate changes.

Stories in magnetism

- As volcanic **lava** solidifies, **igneous rock**s are formed.
- Magnetic materials in the lava line up along the Earth's **magnetic field**.
- The Earth's magnetic field changes over time.
- Geologists can date rocks and track the slow movement of continents using changes in magnetic patterns, linked to radioactive decay.
- This evidence supports **plate tectonic theory**.

Limestone, coal and salt

- Rocks are raw materials found buried in the Earth's crust. Coal, **salt** and limestone are three important raw materials.
- 200 years ago the industrial revolution started in north-west England. Chemical industries built up near to raw materials and transport links.
- There was coal in south Lancashire, salt in Cheshire and limestone in the Peak District. The port of Liverpool and the canal system provided good transport links.

- Limestone formed while Britain was covered by sea:
  - Shellfish died forming **sediments** on the sea bed.
  - Sediments compacted and hardened to form limestone, a **sedimentary rock**.
  - Tectonic plate movements pushed the rock to the surface.
  - Gradually the rocks above were **eroded** away until the limestone was exposed.
- Coal formed in wet swampy conditions when plants like trees and ferns died and became buried. This excluded oxygen, slowing down decay.
- Salt formed while Cheshire was covered by a shallow sea:
  - Rivers brought **dissolved** salts into the sea.
  - Climate warming **evaporated** the water, leaving salt that mixed with sand blown in by the wind.
  - Rock salt formed and was buried by other sediments.

EXAM TIP

Make sure you know how limestone, coal and salt are formed.

- Geologists have found evidence for limestone, coal and salt formation.
- Coal contains fossils of the plants that formed it.
- Limestone contains bits of shell fragments from sea creatures.
- Rock salt contains different-shaped water-eroded grains and wind-eroded grains.
- Ripple marks in rocks indicate water flow from rivers or waves in the sea.

Improve your grade

Limestone, coal and salt

Foundation: Coal, limestone and salt are major raw materials for industry. Choose one and state how it is formed.
AO1 [4 marks]

Salt

Extracting and using salt

- Salt is used in: the food industry; as a source of chemicals; to treat icy roads in winter.
- Salt can be obtained from: collecting and evaporating sea water; mining underground deposits of rock salt.

- Salt is sodium chloride (NaCl) and has many industrial uses.
- Rock salt is spread on icy roads because:
  - the rock is insoluble but the sand in the rock salt gives grip
  - it shows up so people know when roads have been gritted
  - the salt in solution lowers the freezing point, preventing ice forming as easily.
- Only one rock salt mine exists in Britain (in Cheshire). It mines a million tonnes a year.
- If more salt is needed it is usually imported.
- Salt extraction from sea water is only economical in hot climates.
- Purer salt can be obtained by solution mining, which is mainly automatic.

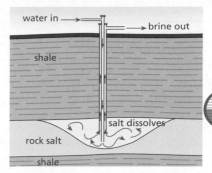

Solution mining – water is pumped at high pressure into the rock salt, the salt dissolves and the salt solution is pushed to the surface.

- Mining rock salt and solution mining can cause subsidence. About half the rock salt cannot be mined, as it is left in place for support.
- Mining can allow water in mines, which may let salt leach out into water supplies, contaminating them.
- Evaporating salt from sea water takes up large areas and spreads salt into the local environment, damaging habitats.

Remember!
Salt is a valuable raw material, but its extraction can have an environmental impact.

The risks of eating salt in foods

- Salt is used in food both as flavouring and as a preservative.
- A higher salt level prevents bacteria growth.
- Too much salt is bad for your health.

- Many people are worried about salt intake, which can cause **high blood pressure**, heart failure and strokes.
- This means salt is classified as a **hazard**.
- A **risk** is the chance of getting ill, and the consequences if you did.
- Risk can be estimated by measuring salt intake.
- Food labels show the amount of salt contained in the product.
- Knowing the risk allows you to make decisions.

Food labels give the amount of salt in the food and the percentage of the recommended daily allowance.

- The government Department of Health (DH) and the Department of the Environment, Food and Rural Affairs (Defra) are responsible for carrying out risk assessment for chemicals in food and advising the public about how food affects health.

EXAM TIP

Make sure you know how to determine risk.

Ideas about science

You should be able to:

- identify, and suggest, examples of unintended impacts of human activity on the environment, such as those resulting from salt extraction
- discuss a given risk, taking account of the chance of it occurring, and the consequences if it did
- discuss the public regulation of risk, e.g. salt levels in food, and explain why it may in some cases be controversial, e.g. whether the daily guideline allowance for salt should be the same for everyone.

Improve your grade

Extracting and using salt

Higher: Describe how salt is obtained by solution mining, and suggest why the chemical industry prefers this method of extraction.   *AO1* [4 marks]

Reacting and making alkalis

About alkalis

- **Alkalis** make **indicators** change colour. Litmus turns blue in alkalis and red in **acids**.
- Alkalis neutralise acids to make **salts**. This is called **neutralisation**.
- The word equation for neutralisation is: acid + alkali → salt + water

Using alkalis

- Alkalis are used for: dyeing cloth; neutralising acid soil; making soap; making glass.
- Stale urine and ash from burnt wood were used in the past as sources of alkalis.
- Due to increased industrialisation, by the 1900s demand for alkalis outstripped the supply.

- In the past, one major use of soap was for cleaning wool. Soap was made by mixing the ashes from burnt wood (called potash) with animal fat and boiling it.
- In coastal areas, seaweed or seaweed ash (called soda) could be used to neutralise acidic soils.
- The first alkali to be manufactured was lime (calcium oxide). This was done by heating limestone (calcium carbonate) in a lime kiln, using coal as fuel.
- Lime is used for: neutralising acidic soils; making glass when heated with sand; removing impurities when iron is made.

- Before modern dyes, clothes were coloured using dyes from plants and animals.
- Alum is a mordant that 'sticks' dye to a fabric. It was purified by reacting it with ammonia contained in stale urine.

Making alkalis

- In 1787 the Frenchman Nicholas Leblanc discovered how to manufacture an alkali.

- The Leblanc process made sodium carbonate by reacting salt and limestone, heated with coal.
- It gave off large amounts of hydrogen chloride (an acidic, harmful gas). It also produced heaps of solid waste, called galligu, that slowly released hydrogen sulfide, a foul-smelling, toxic gas.
- Later, a process was invented to change the harmful hydrogen chloride into useful substances:
  - chlorine used to bleach textiles prior to dyeing
  - hydrochloric acid, which is a starting material for making other chemicals.
- Chlorine can be made by reacting hydrochloric acid and manganese dioxide.
- **Oxidation** converts hydrogen chloride to chlorine.
- **Compounds** have different properties from those of the **elements** they contain.

Remember!
Pollution problems can sometimes be solved by turning waste into useful chemicals.

Patterns of reactions

- An alkali is a solution with a **pH** greater than 7. It turns pH indicator blue or violet.
- Alkalis are **soluble** metal hydroxides and soluble metal carbonates. Some examples are:

Soluble hydroxides	sodium hydroxide – NaOH potassium hydroxide – KOH
	calcium hydroxide – Ca(OH)$_2$
Soluble carbonates	sodium carbonate – Na$_2$CO$_3$ potassium carbonate – K$_2$CO$_3$

- Most metal hydroxides and metal carbonates are insoluble. They are not alkalis but are called **bases**. Bases react with acids in a similar way to alkalis, but do not affect indicators.
- The general pattern of these reactions is: hydroxide + acid → salt + water

  carbonate + acid → salt + water + carbon dioxide gas

- For example: sodium hydroxide + sulfuric acid → sodium sulfate + water
- The salts produced by different acids are:

Acid	hydrochloric – HCl	sulfuric – H$_2$SO$_4$	nitric – HNO$_3$
Salt	chloride – Cl	sulfate – SO$_4$	nitrate – NO$_3$

Higher

Improve your grade

About alkalis

Foundation: Use the tables above to write a word equation for making potassium nitrate.   *AO2* [4 marks]

Uses of chlorine and its electrolysis

The benefits and risks of adding chlorine to drinking water

- In the 19th century, many people died from drinking dirty water.
- Chlorine is now added to water supplies to kill microorganisms.

- The introduction of chlorination made a major contribution to public health.
- Chlorination killed water-borne microorganisms that cause diseases like cholera and typhoid.
- A **correlation** exists between the start of water chlorination in the USA and a fall in the number of deaths from typhoid.

- Chlorine is a toxic gas and can affect human health if too much is present in water.
- Some people disapprove of adding chlorine to water supplies.
- People using mains water supplies have no choice about chlorination.
- Chlorine can react with organic materials in water supplies, forming toxic or carcinogenic compounds called disinfectant by-products (DBPs).
- In the UK the government has decided that the risk from DBPs is very small, so the benefits of disinfecting water outweigh the risks.

Remember!
Chlorination is used throughout UK as the benefits of killing dangerous microorganisms outweigh the possible risk of cancer.

Electrolysis of brine

- **Electrolysis** breaks up compounds using an electric current.
- The electrolysis of brine (sodium chloride solution) makes: chlorine gas; hydrogen gas; sodium hydroxide solution.
- All three products have uses, so there is no waste.

- Electrolysis causes a chemical change, making new products.
- The **anode** is the positive **electrode** and the **cathode** is the negative electrode.
- Large amounts of electricity are needed for electrolysis, so it is expensive.
- The **membrane cell** method is one way to electrolyse brine continuously.
- During brine electrolysis, chlorine forms at the anode and hydrogen at the cathode.
- Industrial uses of these products are:
  - chlorine for making **plastics** like **PVC**, in medicines and crop protection
  - hydrogen for making margarine, as rocket fuel, in fuel cells in vehicles
  - sodium hydroxide for paper recycling, industrial cleaners and refining aluminium.

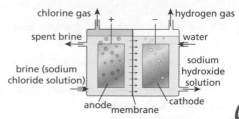

Hydrogen, chlorine and sodium hydroxide are the products of electrolysing sodium chloride solution in the membrane cell.

- Brine electrolysis is one of the most widely used industrial processes.
- While the products have many uses, they can have an environmental impact.
  - Chlorine products, e.g. from fridges and aerosols, are linked to ozone depletion and have been banned.
  - Chlorine used in paper bleaching releases dangerous dioxins, increasing the risk of cancer.
  - The mercury diaphragm method of electrolysis, which is still used, releases mercury waste. This can enter the food chain and is a cumulative poison.
  - Plastics made using chlorine are non-biodegradable.

Ideas about science

You should be able to:
- offer reasons for people's willingness (or reluctance) to accept the risk of a given activity
- explain why it is impossible for anything to be completely risk-free.

Improve your grade

The benefits and risks of adding chlorine to drinking water
Foundation: Worldwide, over 100 000 people die from a disease called cholera each year, through drinking dirty water. In Britain, cholera is rare.

Explain these statements, and suggest advice for people who drink water directly from rivers.

AO2 [4 marks]

Industrial chemicals and LCA

Are chemicals safe?

- Chemicals contain elements. Elements cannot be destroyed, so they remain in the environment forever.
- A risk assessment is used to find out how dangerous substances are.

- Chemicals in the form of solids, liquids and gases can spread out into the environment.
- Some toxic chemicals persist in the environment, can be carried over large distances, and may accumulate in food chains, ending up in human tissues.
- To decide the level of risk of a particular chemical we need to know:
  - how much of it is needed to cause harm
  - how much will be used
  - the chance of it escaping into the environment
  - who or what it may affect.

- Thirty years ago, European laws made risk assessment compulsory for new chemicals.
- Many substances we have used for years have not been tested or there is insufficient data about them for risks to be judged.
- Many people perceive the risk is greater for newer chemicals with less familiar names, while the actual risk could be less.

Should we worry about PVC?

- PVC is a plastic containing carbon, hydrogen and chlorine.
- Small molecules called **plasticisers** are added to PVC to make it softer.
- Plasticised PVC is used to cover electrical wires, for clothing and for seat covers.

- Plasticiser molecules can leach out of PVC into the surroundings, where they may have harmful effects.
- Although chemicals used for plasticisers have passed safety tests, they may have long-term effects on fish, and large amounts have been shown to harm animals.
- As a precaution, plasticised PVC children's toys have been banned in the USA and Europe.

- If PVC is burnt, it gives off toxic gases including dioxins. If eaten, these chemicals build up in fat and are thought to cause cancer.
- Plasticisers are relatively new so long-term studies are not possible. Because of this many people dispute the risks.

Life Cycle Assessment

- A **Life Cycle Assessment** (LCA) measures the energy used to make, use and dispose of a substance, and its environmental impact.

- There are four main stages of an LCA (see diagram), which is sometimes called a 'cradle, use, grave' assessment.
- At each stage of an LCA we need to consider:
  - How much natural resources are required?
  - How much energy is needed or produced?
  - How much water and air is used?
  - How is the environment affected?
- When an LCA has been completed, different products can be compared fairly.

2 Making the product from the chemicals – including transporting the chemicals and the finished product

1 Preparing the chemicals from raw materials found in plants, animals, rocks, the oceans or the air

3 Using the product

4 Disposing of the product and the materials in it when it is of no more use

Stages in the life cycle of a product.

- To produce a fair and **accurate** LCA, a lot of data is required.
- Some aspects are hard to measure, e.g. the lifetime of a car and its disposal method can vary.

Remember!
A Life Cycle Assessment shows the total energy and environmental impact of a product, from getting the raw materials to making it to final disposal (cradle to grave).

Improve your grade

Should we worry about PVC?

Higher: Cling-film is used to wrap food. It may be made from thin sheets of PVC, which contain plasticiser molecules called phthalates to increase flexibility. These molecules have passed safety tests. Despite this, some people are still worried about their safety.

Suggest reasons for some people's reluctance to accept the risk. *AO2* [4 marks]

C3 Summary

Rocks and continents

Geologists look for fossils, shell fragments, grain shape and ripple marks for evidence of conditions when sedimentary rocks form.

Britain is made from rocks that originally came from ancient continents.

The Earth's crust is made very slowly by moving tectonic plates.

Mountain building, erosion, sedimentation, dissolving and evaporation have led to the formation of raw materials.

The climate in Britain has varied from tropical swamps to ice ages.

Salt and alkalis

Rock salt is obtained by mining. Salt is obtained by evaporating sea water and solution mining.

All methods of obtaining salt can have environmental impacts.

Salt is used as a preservative or flavouring in food, but too much can cause health problems.

Government agencies assess food safety and give advice about healthy eating.

Reacting and making alkalis

Alkalis neutralise acid soils, and are used in dyeing clothes, and making soap and glass.

In the past, ash from burnt wood and seaweed were used as alkalis.

Early alkali manufacture produced toxic by-products.

Soluble hydroxides and carbonates are called alkalis. Insoluble ones are bases.

Uses of chlorine and its electrolysis

Adding chlorine to water supplies killed microbes, and this greatly improved public health.

Some people disagree with adding chlorine to drinking water as products might cause cancer, but the benefits outweigh the risk.

Perceived risk and calculated risk are different.

Electrolysis is used to separate sodium chloride solution into hydrogen, chlorine and sodium hydroxide, all of which have many uses.

Electrolysis requires large amounts of electricity, and can have an environmental impact.

Industrial chemicals and Life Cycle Assessment

Small molecules called plasticisers make PVC more flexible.

Plasticisers can leach out and may have harmful effects.

Life Cycle Assessment measures the energy and environmental impact of a product from its cradle to grave.

Notes

Atoms, elements and the Periodic Table

The history of the Periodic Table

- An **element** contains all the same type of **atoms**.

- The modern **Periodic Table** is based on the Russian chemist Dmitri Mendeleev's ideas.

- Mendeleev arranged elements into **groups** (vertical columns) and **periods** (horizontal rows) based on their **relative atomic masses** and patterns in their properties.

- Mendeleev left gaps for undiscovered elements and predicted properties of missing elements.

- Johann Döbereiner noticed 'triads' that linked patterns of the relative atomic masses for three elements.

- John Newlands noticed an 'octaves' pattern, where every eighth element had similar properties.

- Scientists rejected Döbereiner's triads and Newlands' octaves because most elements did not fit their 'patterns'.

- When new elements were discovered, they fitted Mendeleev's predictions.

- Data about properties of elements in the Periodic Table can be used to work out trends and to make predictions.

Lines of discovery

- When elements are heated they emit coloured flames. Some elements emit distinctive flame colours, e.g. lithium salts produce a red flame.

- The coloured light can be split into a **line spectrum** that is unique to each element.

- The discovery of some of elements, e.g. helium, happened because of the development of spectroscopy.

- Helium was discovered when chemists looked at the line spectrum from the sun.

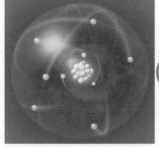

The line spectrum of sodium is two single yellow lines, so close together they look like one.

Remember!
Every element has a different pattern of lines. The pattern of lines in a line spectrum can be used to identify the element.

Inside the atom

- Atoms have a tiny, central nucleus that contains **protons** and **neutrons**.

- **Electrons** travel around the outside of the atom in **shells**.

Particle	Charge	Mass
proton	+1	1
electron	−1	almost zero
neutron	0	1

- All the atoms of a particular element have the same number of protons.

- Atoms have the same number of protons and electrons.

- The modern Periodic Table arranges atoms in order of their **proton number**.

- Number of protons + number of neutrons = relative atomic mass

A diagram of an atom showing the electrons orbiting the nucleus, which is made up of protons and neutrons (the nucleus is not to scale).

 Higher

● Ideas about science

You should be able to:

- separate data from explanations about the Periodic Table

- discuss why Mendeleev's table was an improvement on earlier ideas

- discuss how important it was that Mendeleev's predictions about elements were correct

- use data about elements to identify trends in properties and make predictions using the Periodic Table.

● Improve your grade

The history of the Periodic Table

Foundation: Explain why Mendeleev's arrangement of elements was an improvement on Döbereiner's triads and Newlands' octaves.
AO1 [4 marks]

Electrons and the Periodic Table

Sorting electrons

- Electrons are arranged in shells around the nucleus.
- The first shell is closest to the nucleus and can hold 2 electrons.
- The second and third shells are further away from the nucleus.
- The second shell holds 8 electrons.
- The **electron arrangement** of oxygen can be written 2.6.
- The electron arrangement of chlorine can be written 2.8.7.

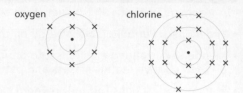

Electron arrangements of oxygen and chlorine atoms. Each cross represents an electron.

- The number of electrons in an atom is the same as the number of protons.
- The number of electrons in an atom is the same as the proton number.
- For the first 20 elements, the third shell holds 8 electrons.

Electron shell	Number of electrons
1	2
2	8
3	8

- Potassium, proton number 19, has an electron arrangement 2.8.8.1.

- Electrons in different shells have different **energy levels**.
- The closer the electron shell is to the nucleus, the lower the energy level.

Magnesium follows sodium in the Periodic Table, and has one extra proton and one extra electron in the third shell.

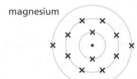

Finding elements in the Periodic Table

- In the Periodic Table, elements are arranged in order of proton number.
- A row across the Periodic Table is called a period.
- The number of electrons in the outer shell increases across a period.
- As you move from left to right along a period, elements change from **metals** to non-metals.

EXAM TIP

Make sure you know how to find the proton number of an atom from the Periodic Table. You need to be able to use the proton number to work out the electron structure of an atom by drawing 'dot and cross diagrams' (like the ones for oxygen and chlorine on this page) and by writing electron arrangements, e.g. 2.6 or 2.8.7.

- As you move from left to right along a period, each element has one more proton and one more electron.
- Properties, e.g. melting points, change across a period. These changes are called **trends**.
- Elements in Group 0 have full electron shells and are **inert** – this means they are very unreactive.

- The number of electrons in the outer shell of an atom is the same as its group number on the Periodic Table.
- Atoms of elements with up to 3 electrons in their outer shell are metals.
- Atoms of elements with 5 or more electrons in their outer shell are non-metals.
- Elements with full outer shells are the inert gases.

Improve your grade

Finding elements in the Periodic Table

Higher: An atom has the electronic arrangement 2.8.1.

Identify the element and explain why its electronic arrangement shows that it is likely to be a metal.

AO2 [3 marks]

Reactions of Group 1

Group 1 – the alkali metals

- A **group** is a vertical column in the Periodic Table.
- Group 1 is called the **alkali metals**.

- All elements in Group 1 are metals and have one electron in the outer shell of their atoms.
- Group 1 elements are soft metals that can be cut with a knife. The freshly cut surface is shiny but it **tarnishes** quickly in moist air by reacting with oxygen.
- The physical properties of Group 1, e.g. melting point, boiling point and density, show trends down the group.

- The reactivity of Group 1 elements is linked to the single electron in the outer shell. They all form **ions** with a 1+ charge by losing an electron.
- The outer electron is easiest to lose if the atom is bigger (because the electron is further from the nucleus). So reactivity increases down the Group 1 elements as the atoms get bigger.

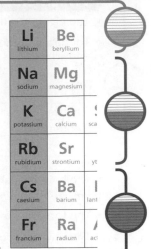

Li	Be
lithium	beryllium
Na	Mg
sodium	magnesium
K	Ca
potassium	calcium
Rb	Sr
rubidium	strontium
Cs	Ba
caesium	barium
Fr	Ra
francium	radium

Group 1 in the Periodic Table – the alkali metals.

Reactions of Group 1 elements with water

- Group 1 elements all react with water, e.g. lithium and sodium fizz and move around on the surface of the water. The reaction gets more violent as you move down Group 1, e.g. potassium explodes and rubidium explodes more violently.
- In the reaction, hydrogen gas is formed (which 'pops' when lit).
- The reaction also makes a metal hydroxide, which is an alkali and turns **pH** indicator blue.
- Group 1 metals are flammable and their hydroxides are harmful and corrosive. When handling Group 1 metals they should be kept away from water and naked flames.

highly flammable   toxic   corrosive

harmful   explosive   oxidising

Take precautions when you see these hazard symbols.

- The equation for the reaction of Group 1 elements with water is:
  metal + water → metal hydroxide + hydrogen
- If we use M to stand for any of the Group 1 metals, the general equation is:
  $2M(s) + 2H_2O(l) \longrightarrow H_2(g) + 2MOH(aq)$
- For example, the **balanced symbol equation** for the reaction of sodium with water is:
  $2Na(s) + 2H_2O(l) \longrightarrow H_2(g) + 2NaOH(aq)$
- For a balanced equation, the numbers of atoms of each element must be the same on both sides of the equation.

EXAM TIP

Exam questions often ask you to write or complete equations for the reactions of Group 1 metals with water and chlorine. Practise writing them. Make sure that you know the **formulae** of the compound and what numbers to use to balance the equations.

Reactions of Group 1 elements with chlorine

- Sodium reacts vigorously with chlorine to give a yellow flame; it makes a white solid (sodium chloride).
- The other Group 1 metals react in a similar way, and the reactions are faster down the group.

- The word and symbol equations for the reaction between sodium and chlorine are:
  sodium + chlorine → sodium chloride
  $2Na(s) + Cl_2(g) \longrightarrow 2NaCl(s)$

Remember!
Group 1 elements show a trend in reactivity. They get more reactive 'down the group'.

Improve your grade

Reactions of Group 1 elements with chlorine
Higher: Write word and symbol equations for the reaction of sodium with bromine. Compare the rate of reaction of sodium and potassium with bromine.

AO1 [3 marks]

Group 7 – the halogens

Which elements are halogens?

- The elements in Group 7 are called the halogens.
- The table shows the appearance of the halogens at room temperature and when they are warmed to form gases.

Halogen	Appearance at room temperature	Colour of gas
chlorine	pale green gas	pale green
bromine	red-brown liquid	reddish-brown
iodine	dark grey solid	purple

		He helium
O oxygen	F fluorine	Ne neon
S sulphur	Cl chlorine	Ar argon
Se selenium	Br bromine	Kr krypton
Te tellurium	I iodine	Xe xenon
Po polonium	At astatine	Rn radon

Group 7 of the Periodic Table, the halogens.

- The halogens all contain **diatomic molecules**. This means that they have two atoms joined together in each molecule.
- The formulae of the halogens are: chlorine: Cl_2; bromine: Br_2; iodine: I_2.
- The physical properties of the halogens show a trend down the group, e.g. melting points and boiling points increase.

Patterns in Group 7

- Group 7 elements are **corrosive** and **toxic**. They need to be used in a fume cupboard.
- Group 7 elements react with alkali metals and with other metals such as iron to form metal **halides**. Some examples are shown in the table.

Metal		Halogen		Metal halide
iron	+	chlorine	→	iron chloride
sodium	+	bromine	→	sodium bromide
potassium	+	iodine	→	potassium iodide

- Halogens are less reactive down the group. For example:
  - Sodium reacts vigorously in chlorine but reacts less violently with iodine.
  - Iron reacts vigorously in contact with fluorine but only reacts with iodine when heated.
  - Fluorine reacts even more vigorously than the other halogens.
- **Displacement reactions** happen when a more reactive halogen takes the place of a less reactive halogen in a compound.
  - Chlorine is more reactive than bromine and displaces bromine from potassium bromide solution.
  - Bromine is less reactive than chlorine and cannot displace chlorine from potassium chloride solution.

Remember!
The reactivity of Group 7 decreases (gets slower) down the group – this is the opposite pattern to Group 1.

When chlorine water (a solution of chlorine in water) is added to potassium bromide solution, the chlorine displaces orange-coloured bromine.

- All halogen atoms have 7 electrons in their outer shell.
- The trend in reactivity is linked to the number of electron shells in the atom. For non-metals, the smaller the atom (the fewer the electron shells) the more reactive the element.

Improve your grade

Patterns in Group 7

Higher: Liz adds chlorine water to potassium bromide solution. The table shows what she sees and her explanation.

Halogen	Compound	Observations halide	Explanation
Chlorine	Potassium bromide	Solution turns brown	Bromine is made because chlorine displaces bromine. Chlorine is more reactive than bromine.

Predict what you will see when chlorine water is added to potassium iodide solution. Explain your reasoning.

AO2 [4 marks]

Ionic compounds

What are ionic compounds?

- Compounds of a Group 1 element and a Group 7 element (e.g. sodium chloride) are solids with high melting points.

- These compounds are **ionic compounds** because they contain charged particles, or ions, that are arranged in a regular pattern called a **crystal lattice**.

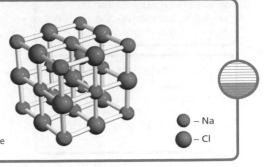

The ions in sodium chloride are arranged in a crystal lattice.

– Na
– Cl

Explaining properties

- Ionic compounds are **soluble** in water and conduct electricity when they are melted.

- When an ionic crystal melts or **dissolves** in water the ions are free to move.

- An ionic compound conducts electricity when molten or in solution but not when solid. This is because the ions must move to the **electrodes** to complete the circuit.

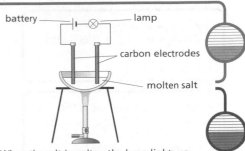

battery — lamp
carbon electrodes
molten salt

When the salt is molten the lamp lights up, showing that a current is passing through it.

Forming ions, formulae and charges

- An atom of a Group 1 element *loses* one electron to become a *positive* ion, e.g. Na becomes Na^+.

- An atom of a Group 7 element *gains* an electron to become a *negative* ion, e.g. Cl becomes Cl^-.

- When a Group 1 metal atom becomes an ion it loses one electron from its outer shell.

- All Group 1 metals become ions with a 1+ charge.

- When a Group 7 metal atom becomes an ion it gains one electron in its outer shell.

- All Group 7 metals become ions with a 1− charge.

- The ions have the same electron arrangement as an atom in Group 0. For example:

Electron arrangement	
atom	ion
Na: 2.8.1	Na^+: 2.8
Cl: 2.8.7	Cl^-: 2.8.8

same electron arrangement as a neon atom 2.8
same electron arrangement as an argon atom 2.8.8

- In the formula for an ionic compound the number of positive and negative charges balance, e.g. Na^+ and Cl^- have one positive and one negative charge and so form the compound NaCl.

- You can work backwards from the formula to work out the charge on the ions, e.g. in $CaCl_2$, there are 2− charges from the two Cl^- ions, so the calcium ion has a charge of Ca^{2+}.

sodium atom
Na
2,8,1

sodium ion
Na^+
2,8

chlorine atom
Cl
2,8,7

chlorine ion
Cl^-
2,8,8

EXAM TIP

Some of the formulae of compounds have more than one atom grouped together into an ion. The metal hydroxides in this topic contain hydroxide ions OH^-. Make sure you know their formula and charge – you might be asked to work out the formula of an unfamiliar hydroxide in the exam.

Higher

Improve your grade

Explaining properties

Foundation: Explain why sodium chloride conducts electricity when it is molten or dissolved in water but not when solid.

AO1 [4 marks]

C4 Summary

Many scientists worked to organise elements into patterns based on their properties.

Atoms contain protons, neutrons and electrons. The numbers of each can be worked out using the proton number and atomic mass of the element.

Atoms, elements and the Periodic Table

Mendeleev's ideas formed the basis for the modern Periodic Table.

Each element has a unique flame-test colour and line spectrum.

Electrons are arranged in shells around the nucleus and can be shown using electron arrangements.

A period is a row across the Periodic Table. There are trends in properties from left to right, e.g. metals to non-metals.

Electrons and the Periodic Table

Each shell holds a maximum number of electrons.

The position of an element on the Periodic Table gives information about its properties and its electron arrangement.

Group 1 metals are all soft, can be cut with a knife and tarnish in moist air.

The reactivity of Group 1 metals increases down the group and is linked to the electron arrangement of the atoms.

Reactions of Group 1

Group 1 metals react violently with water to give hydrogen and a metal hydroxide.

Group 1 metals react with chlorine to give a metal chloride.

Halogens react with metals and other less reactive halogen compounds (displacement reactions).

The appearances of the halogens are all different at room temperature and as gases. They all contain diatomic molecules, e.g. Cl_2.

Group 7 – the halogens

The halogens get less reactive down the group and this is linked to the electron arrangement of the atoms.

Positive ions form when atoms lose outer-shell electrons.

Negative ions form when they gain outer-shell electrons.

Ions have the same electron arrangement as Group 0 gases.

Ionic compounds

Ionic compounds conduct electricity when molten or in solution because the ions can move.

The formula of an ionic compound can be worked out by making sure that there are enough positive and negative ions so that the total charges balance.

Molecules in the air

Air

- **Dry air** contains non-metal elements, e.g. nitrogen, oxygen and argon.

- Air also contains small amounts of non-metal compounds, e.g. carbon dioxide and water vapour.

- The gases are non-metal elements and non-metal compounds. They are molecular because they contain atoms joined together into small molecules.

- The formulae and amounts of some of the gases in the air are shown in the table.

Substance	Formula	Percentage of dry air
nitrogen	N_2	78
oxygen	O_2	21
argon	Ar	1
carbon dioxide	CO_2	0.04

- Molecules can be shown either in 2-D or in 3-D.

- The atoms in the molecules are held together by **covalent bonds**.

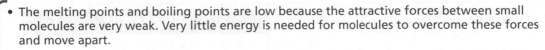

2-D and 3-D diagrams of some simple molecules.

- A covalent bond forms when atoms share a pair of electrons.

- The atoms are held together because the positively charged nuclei of both atoms are attracted to the negatively charged shared pair of electrons.

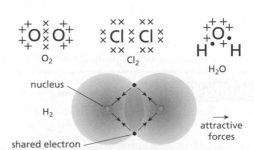

Simple molecular substances

- Simple molecular substances, such as molecules in the air, have very low melting points and boiling points.

Substance	Melting point (°C)	Boiling point (°C)
nitrogen	−210	−196
oxygen	−218	−183

> ### EXAM TIP
> A common question is to ask about data relating to melting points and boiling points of gases. Any substance with a melting point and boiling point below 25 °C is a gas at room temperature. Remember that many values for gases are negative. Bigger negative numbers mean lower boiling points and melting points.

- The melting points and boiling points are low because the attractive forces between small molecules are very weak. Very little energy is needed for molecules to overcome these forces and move apart.

- Molecules of elements and compounds have no electrical charge, so pure molecular substances cannot conduct electricity.

- For small covalent molecules, the forces *between* molecules are weak, but the forces *within* molecules (covalent bonds) are strong.

- When a molecular substance melts, the molecules are easily separated from one another but the molecules themselves are not broken up into separate atoms.

Improve your grade

Simple molecular substances

Higher: The table shows some data about oxygen.

Boiling point	State at room temperature	Density	Electrical conductivity
−218 °C	gas	very low	does not conduct

Use ideas about bonding and forces between molecules to explain the properties of oxygen.

AO2 [3 marks]

Ionic compounds: crystals and tests

Ionic crystals

- The Earth's **hydrosphere** is all the water on Earth, including oceans, seas, lakes and rivers.
- The hydrosphere is mostly water, with some dissolved compounds called **salts**.
- When water evaporates, dissolved salts form solid **crystals**.
- Salts are **ionic compounds**. **Ions** have either a positive or a negative charge and are arranged in a giant 3-D pattern called a **lattice**.

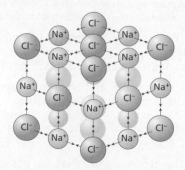

- The strong force of attraction between positively charged and negatively charged ions is called an **ionic bond**.
- Ionic compounds have high melting and boiling points because a large amount of energy is needed to overcome the forces between ions in the lattice.
- Ionic compounds do not conduct electricity when solid, because the ions are not free to move. When they are melted or dissolved in water, the ions can move and they conduct electricity.

The forces between positively and negatively charged ions pulls them together in a giant 3-D crystal lattice.

Higher

- In the formula for an ionic compound, the number of positive charges just balance and cancel out the negative charges, e.g. Na^+ and Cl^- make $NaCl$; Mg^{2+} and Cl^- make $MgCl_2$.
- Some ions, e.g. the sulfate ion SO_4^{2-}, contain groups of atoms. This is called a **molecular ion**.

Positively charged ions	Negatively charged ions
sodium (Na^+)	chloride (Cl^-)
potassium (K^+)	bromide (Br^-)
magnesium (Mg^{2+})	iodide (I^-)
calcium (Ca^{2+})	sulfate (SO_4^{2-})

Testing for ions

- Ions in a compound can be identified by their distinctive properties, e.g. compounds containing the copper ion are often blue.
- Solutions of some ionic compounds make a **precipitate** of an **insoluble** compound when they mix. The colour of the precipitate can be used to identify the ions in the compound.
- Adding an **alkali**, such as dilute sodium hydroxide, to different positive metal ions gives different colours of precipitate, as the table shows.

Ion	Observation
calcium, Ca^{2+}	White precipitate (insoluble in excess)
copper, Cu^{2+}	Light blue precipitate (insoluble in excess)
iron(II), Fe^{2+}	Green precipitate (insoluble in excess)
iron(III), Fe^{3+}	Red-brown precipitate (insoluble in excess)
zinc, Zn^{2+}	White precipitate (soluble in excess, giving a colourless solution)

- Negative carbonate ions are identified by adding dilute acid and looking for fizzing (**effervescence**).
- Negative chloride, bromide, iodide and sulfate ions are identified by adding dilute silver nitrate or dilute barium chloride and looking for precipitates.

EXAM TIP

You are not expected to know these tests 'off by heart'. In the exam you will have a sheet of tests and expected results to refer to, like the one on page 84. Practise using the sheet so it is familiar to you.

- Carbonates fizz when an acid is added because carbon dioxide gas is made in the reaction.
- Precipitates form when an insoluble solid is made in the reaction. For example, most metal hydroxides are insoluble, as is silver chloride from the reaction of silver ions and chloride ions.

Higher

- **Ionic equations** with **state symbols** show what happens when precipitates form. For example:
  - For positive ions reacting with hydroxide ions: $Cu^{2+}(aq) + 2OH^-(aq) \rightarrow Cu(OH)_2(s)$
  - For negative ions reacting with silver ions or barium ions: $Ag^+(aq) + Cl^-(aq) \rightarrow AgCl(s)$

Improve your grade

Testing for ions

Foundation: Sam tests a salt. He finds out it contains copper ions.

Describe what Sam does and what he sees.

AO2 [3 marks]

Giant molecules and metals

Metals, minerals and ores

- The **lithosphere** is the rigid outer layer of the Earth, made up of the crust and upper mantle. It contains rocks and **minerals**.

- Minerals are solids with atoms or ions arranged in a regular arrangement or lattice, e.g. carbon in the form of diamond or graphite.

- Silicon, oxygen and aluminium are very abundant elements in the Earth's lithosphere.

- Most of the silicon and oxygen on Earth are joined together in the Earth's crust as silicon dioxide, for example in the mineral quartz.

- Some minerals contain metals. Rocks that contain metal minerals are called **ores**.

- Some ores contain metal oxides.

- Copper, zinc and iron can be extracted from their ores by heating their metal oxides with carbon. Carbon **reduces** the metal oxide by taking away oxygen.

- The amount of minerals in ores varies. For some metals, e.g. copper, a huge amount of rock has to be mined to extract a small amount of metal.

- Extracting metals by heating their oxides with carbon is a **redox reaction**, because both oxidation and reduction happen.

- **Reduction** happens because the metal oxide *loses* oxygen.

- **Oxidation** happens because the carbon *gains* oxygen.

- For example, the extraction of zinc uses carbon to reduce zinc oxide to zinc:

  zinc oxide + carbon ⟶ zinc + carbon dioxide

  $2ZnO(s)$ + $C(s)$ ⟶ $2Zn(s)$ + $CO_2(g)$

- In this reaction the zinc has been *reduced* and the carbon has been **oxidised**.

EXAM TIP

If the question asks for state symbols, you need to decide whether each substance is:
- a solid (s)
- a liquid (l)
- a gas (g)
- dissolved in water (aq).

The state symbols always go after the formula.

Giant covalent structures

- Diamond and graphite contain many carbon atoms covalently bonded together in a regular pattern. This is called a **giant covalent structure**.

- Covalent bonds form when atoms share electrons.

- The bonds in diamond are very strong and need a large amount of energy to break them. This is why diamond has a very high melting and boiling points and does not dissolve in water.

- There are no free charged particles in diamond, so it does not conduct electricity when solid or when melted.

- Silicon dioxide also has a giant covalent structure, so it has similar properties to diamond.

- In diamond, each carbon atom is covalently bonded to four other atoms in a tetrahedral 3-D lattice.

- In graphite, each atom is strongly bonded to three others in sheets. The sheets are strong but there are is only a weak force between the layers, so they can slide over each other.

- There are free-moving electrons between the layers in graphite, so it conducts electricity.

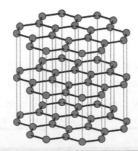

The structure of graphite. Each carbon atom is joined to three others by covalent bonds.

Remember!
Ionic compounds contain charged particles and so conduct electricity when they are molten. Covalent compounds have no charged particles and cannot conduct electricity. Graphite is the only exception to this rule.

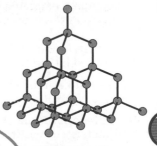

The structure of diamond. Each carbon atom is joined to four others by covalent bonds.

Improve your grade

Giant covalent structures

Higher: Diamond is the hardest naturally occurring material on Earth.

Use ideas about the structure of diamond to explain why.

AO1 [3 marks]

Equations, masses and electrolysis

Equations

- Equations show the chemicals that react together (the reactants) and the chemicals that are made (the products).

- A balanced equation also shows the number of atoms of each element on each side of the equation. Equations are balanced because there is the same number of atoms of each element on both sides.

- State symbols show whether each chemical is solid (s), liquid (l), gas (g) or dissolved in water (aq).

- To balance equations, write the symbols for the elements and the formulae for the compounds. Write numbers in front of the formulae to give the correct number of atoms on each side.

$$2ZnO(s) + C(s) \rightarrow 2Zn(s) = CO_2(g)$$

Representing the balanced equation for the reaction zinc oxide with carbon.

Atomic mass and formula mass

- The **relative atomic mass** of an atom is the mass of an atom compared to the mass of an atom of carbon, which is given the value 12.

- The **relative formula mass** of a compound is the sum of the relative atomic masses of all the atoms or ions shown in its formula.

- For example, to find the relative formula mass of water, H_2O, use the relative atomic masses of hydrogen (1) and oxygen (16). The relative formula mass of water, H_2O is: $(2 \times 1) + 16 = 18$.

- The **gram formula mass** of an element or compound is its relative atomic mass or relative formula mass in grams, e.g. the gram formula mass of water is 18 g.

> **Remember!**
> The atomic mass of every element is in the Periodic Table – find any masses you need by looking there.

- Use this method to work out the percentage of a metal in a mineral:

$$\text{Percentage of metal in mineral} = \frac{\text{total mass of metal atoms}}{\text{gram formula mass}} \times 100\%$$

- To calculate the mass of a metal that can be extracted from a mineral, you need to work out the mass of metal in grams that can be extracted from 1 gram formula mass.

- For example, how much iron can be extracted from 80 g Fe_2O_3? One Fe_2O_3 formula unit can make 2Fe gram formula masses: 160 g can make $(2 \times 56) = 112$ g. So, 80 g can make 56 g.

Using electrolysis

- **Electrolysis** means passing an electric current through an ionic compound when it is either molten or dissolved in water.

- The compound is called the **electrolyte** because it conducts electricity.

- Electrolytes break down, or **decompose**, as the electricity passes through.

- Aluminium is extracted from aluminium oxide by electrolysis. Aluminium and oxygen are made in this process.

- Electrolysis is used to extract more reactive metals (e.g. aluminium) because their oxides cannot be reduced by carbon.

- Metals form at the negative **electrode** because positive metal ions are attracted to the negatively charged electrode (**cathode**).

- Negative ions such as chloride (Cl^-) and oxide (O^{2-}) move to the positively charged electrode (**anode**). Non-metals, such as chlorine and oxygen, form at the anode.

- At the negative electrode, metals ions gain electrons and become neutral metal atoms, e.g. aluminium ions gain 3 electrons to form aluminium atoms: $Al^{3+} + 3e^- \rightarrow Al$

- At the positive electrode, non-metal ions lose electrons and become neutral non-metal atoms, e.g. 2 oxygen ions lose 2 electrons each to form oxygen gas: $2O^{2-} \rightarrow O_2 + 4e^-$

Improve your grade

Using electrolysis

Higher: An electric current is passed through molten potassium chloride.

Write ionic equations to help you to explain what happens at each electrode and name the products that form.

AO1 [5 marks]

Metals and the environment

Metals

- Metals are very useful because they are strong, **malleable** (they can be hammered into shape), have high melting points and are good conductors of electricity.

- Atoms in metals are held together by **metallic bonds**. The atoms are arranged in a regular pattern in a giant lattice.

- Metallic bonds are strong so a lot of energy is needed to melt or reshape them.

- Metal atoms lose their outer-shell electrons to form positive ions. In solid metals, the outer-shell electrons form a 'sea of electrons', which can move freely.

- The attraction between the positive ions and the sea of electrons is very strong, so that metals have high melting points and high strength.

- Metals conduct electricity because the electrons can move.

- In pure metals, all the atoms are the same size and can roll over each other. This means that metals can be reshaped (they are malleable) even though the bonds are strong.

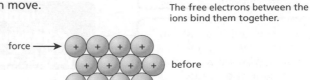

The free electrons between the ions bind them together.

Remember!
Metals and ionic compounds conduct electricity in different ways. Ionic compounds only conduct when they are molten or in solution because the ions need to move.

Applying a force to change the shape of a metal does not affect the arrangement of the atoms, so it remains strong.

before

after

Higher

Metals in the environment

- Some metals are poisonous, e.g. lead, mercury and cadmium.

- Waste poisonous metals from mines destroys habitats and damages soil and water sources.

- Extracting metals makes pollutant gases that cause acid rain.

- Large amounts of waste rock need to be processed to produce very small amounts of some metals such as copper.

- Processing large amounts of rock uses a lot of energy.

- Some minerals contain compounds of metals with sulfur. During extraction of the metal, sulfur dioxide is made. Sulfur dioxide gas forms acid rain and damages plants and fish.

- We need large amounts of copper for electrical wiring and circuits, pipes and building materials. Some waste copper is recycled.

- Lead is a **toxic** metal that was used to make batteries for vehicles.

- Modern batteries are made using lithium. Lithium is not toxic but it is difficult to extract enough lithium to meet the demand for batteries.

EXAM TIP

You do not need to learn facts about individual metals and their effects on the environment, but it is important that you can discuss data and information that you are given in questions. Typical questions will ask you about the costs and benefits of extracting metals – you need to be able to give the benefits of all of the uses of metals to people, as well as discussing the harm that extraction causes to the environment.

Ideas about science

You should be able to:

- discuss the costs and benefits to people and the environment of extracting metals.

- explain what sustainability means and apply ideas about sustainability to extracting metals

- discuss why people may have different views about the impact of mining on their lives and the environment.

Improve your grade

Metals in the environment

Foundation: Old car batteries are made from lead. Lead is toxic. Modern batteries use alternative, non-toxic metals.

Use ideas about cost and benefit to explain why manufacturers could not stop using lead to make batteries until alternative batteries were developed.

AO3 [3 marks]

C5 Summary

Dry air contains 78% nitrogen, 21% oxygen and about 1% argon and other gases such as carbon dioxide.

Gases in the air contain small molecules. The atoms in the molecules are held together by covalent bonds.

The melting points and boiling points of gases in the air can be explained by using ideas about weak forces between their molecules.

Molecules in the air

Ionic salts are compounds containing positive and negative ions in a 3-D lattice.

The negative ions in a salt can be identified by testing with a dilute acid, dilute silver nitrate or dilute barium chloride.

Ionic compounds: crystals and tests

Ionic compounds have very high melting points and conduct electricity when they are molten or dissolved in water but not when they are solids.

The positive ions in a salt can be identified by adding dilute sodium hydroxide and looking at the colour of the precipitate.

The lithosphere contains rocks and useful minerals. Metal minerals are called ores.

The properties of giant covalent structures are related to the way that the atoms are held in strong, 3-D lattices of covalent bonds.

Giant molecules and metals

Metals can be extracted from ores by reactions that involve reduction and oxidation.

Diamond, graphite and silicon dioxide all have giant covalent structures.

Balanced equations show the numbers of atoms that react. The same numbers of atoms are on both sides.

Ionic equations show how metal ions gain electrons at the negative electrode and non-metal ions lose electrons at the positive electrode.

Equations, masses and electrolysis

Relative atomic mass, relative formula mass and gram formula mass can be used to work out how much metal can be extracted from a mineral.

Electrolysis is used to extract reactive metals from their compounds. The compound breaks down when an electric current passes through.

The properties of metals make them useful for a wide range of purposes.

Metals and the environment

Metal extraction causes environmental harm because large amounts of waste rock are produced and some metals are very toxic.

Metals contain positive ions in a sea of free-moving electrons. This structure gives metals high melting points, high strength, malleability and good electrical conductivity.

Making chemicals, acids and alkalis

Making chemicals

- Hazard symbols (see page 25) are used to show that chemicals are hazardous.

- Chemists and engineers must assess the risks before using chemicals to make a new product.

- **Chemical synthesis** means using simple substances to make new, useful chemical compounds.

- The chemical industry uses chemical synthesis to make chemicals for food additives, fertilisers, dyes, paints, pigments and pharmaceuticals (medicines).

Acids and alkalis

- **Indicators** turn different colours in **acids** and **alkalis**. Litmus is red in acids and blue in alkalis. Universal indicator is orange or red in acids and green to blue in alkalis.

- **Pure** acid compounds can be solids, liquids or gases. These compounds dissolve in water to form dilute acids that can be tested using indicators.

State when pure	solid	liquid	gas
Examples of acid	citric acid; tartaric acid	sulfuric acid; nitric acid; ethanoic acid	hydrochloric acid

- Sodium hydroxide, potassium hydroxide and calcium hydroxide are common alkalis.

- The **pH** scale is a measure of how strong an acid or an alkali is.

- The pH can be measured using universal indicator or a pH meter. The colour of the universal indicator can be compared to a colour chart to find the pH number of a sample.

- Neutral solutions have a pH of 7. pH numbers for acids are below 7 and for alkalis are above 7.

Remember!
The lower the pH of an acid, the stronger the acid. The higher the pH of an alkali, the stronger the alkali.

Universal indicator shows a range of colours across the pH scale from 1 (red) to 14 (dark blue).

Reactions of acids

- Acids react with many metals and metal compounds to make a **salt**.

- Acids react with many metals to form a salt and hydrogen gas, e.g.:
  calcium + hydrochloric acid → calcium chloride + hydrogen

- Acids react with metal oxides and hydroxides to form a salt and water, e.g.:
  magnesium oxide + sulfuric acid → magnesium sulfate + water
  sodium hydroxide + nitric acid → sodium nitrate + water

- Acids react with metal carbonates to form a salt, water and carbon dioxide gas, e.g.:
  calcium carbonate + hydrochloric acid → calcium chloride + water + carbon dioxide

- Salts are **ionic compounds** and contain a positively charged metal **ion** and a negative ion from the acid.

Acid	hydrochloric	sulfuric	nitric
Formula	HCl	H_2SO_4	HNO_3
Negative ion	chloride (Cl^-)	sulfate (SO_4^{2-})	nitrate (NO_3^-)

- The reactions of acids can be shown using symbol equations with **state symbols**.

- The state symbols can be either solid (s), liquid (l), gas (g) or solution in water (aq), e.g.:
  – $Ca(s) + 2HCl(aq) \rightarrow CaCl_2(aq) + H_2(g)$ – $Mg(OH)2 + H_2SO_4(aq) \rightarrow MgSO_4(aq) + H_2O(l)$
  – $NaOH(aq) + HNO_3(aq) \rightarrow NaNO_3(aq) + H_2O(l)$ – $CaCO_3(s) + 2HCl(aq) \rightarrow CaCl_2(aq) + H_2O(l) + CO_2(g)$

- To work out the formula of a salt, the number of positive charges must equal the number of negative charges. For example, in potassium sulfate, two potassium ions (K^+) are needed to balance the charge on the sulfate ion (SO_4^{2-}), so the formula is K_2SO_4.

Metal	sodium	potassium	magnesium	calcium	copper	zinc	iron(II)	aluminium	iron(III)
Ion	Na^+	K^+	Mg^{2+}	Ca^{2+}	Cu^{2+}	Zn^{2+}	Fe^{2+}	Al^{3+}	Fe^{3+}

- A balanced equation has the same number of each type of atom on each side of the equation.

Improve your grade

Reactions of acids

Higher: Write a word equation and a balanced symbol equation to show the reaction between sodium hydroxide and sulfuric acid.

AO1 [3 marks]

Reacting amounts and titrations

Reacting amounts

- The **formula** of a compound is the simplest ratio of the numbers of atoms that are in the compound.

- The **relative atomic masses** of atoms can be found on the **Periodic Table**.

- The **relative formula mass** of a compound is the sum of the relative atomic masses of all the atoms in the formula. For example, the relative formula mass of magnesium chloride, $MgCl_2$, is worked out like this:

  RAM Mg = 24; RAM Cl = 35.5

  Relative formula mass $MgCl_2$ = 24 + (2 × 35.5) = 96

> The relative atomic mass is the number above the symbol of the element. This number compares the mass of the atom with the mass of a carbon atom, which has the value 12.

	fluorine 9	
,2	35.5	4,
S	Cl	A
phur	chlorine	argc
6	17	1
	80	

- A balanced equation shows the atoms and molecules in a reaction. The number of atoms of each type of element is the same on both sides of the equation.

- Relative atomic mass (for atoms) and relative formula mass (for compounds) can be used to work out the amounts of **reactants** and **products** in the reaction.

EXAM TIP

Only work out the relative formula masses for the chemicals mentioned in the equation. In the example below you can ignore the HCl and the H_2O because the question does not ask about them.

- The balanced equation for a reaction can be used to calculate the minimum quantity of reactants needed to make a particular amount of a product.

- For example, what mass of magnesium chloride can be made by reacting 1 tonne of magnesium oxide with hydrochloric acid?

	MgO	+ HCl ⟶ MgCl₂	+ H₂O
RFMs	MgO = 24 + 16 = 40	MgCl₂ = 24 + (2 × 35.5) = 96	
Reacting masses	40 tonnes	96 tonnes	
	1 tonne	96/40 = 2.4 tonnes	

Titrations

- When an acid reacts with an alkali it becomes **neutral**. This is called a **neutralisation** reaction.

- A **titration** is used to measure the volume of acid and alkali that exactly react together.

- An indicator is added so that you can see when neutralisation happens. The indicator suddenly changes colour at the **end-point** of the titration.

Carrying out a titration. The end-point is reached when one drop from the burette changes the colour of the indicator.

- A titration is repeated to check that the results are close together. Variations between readings are small differences that happen due to small experimental errors.

- The **range** of readings is the spread of readings from the highest to the lowest.

- The **true value** should fall within the range of the readings.

- An estimate of the true value can be worked out by calculating a **mean** (an average) of the results.

- An **outlier** is a reading that is very different to most of the others. Outliers should be left out when you calculate the mean, because they are usually the result of errors in measurement.

- The volumes obtained from a titration will always be in the same proportion if the same concentrations of the same solutions are used.

Ideas about science

You should be able to:

- work out a best estimate for a titration by working out the mean of the accurate results

- explain why repeating measurements leads to a better estimate from a set of repeated titration measurements

- make a sensible suggestion about the range within which the true titration value probably lies

- identify any outliers in data from titrations.

Improve your grade

Reacting amounts

Foundation: What is the relative formula mass of calcium hydroxide, $Ca(OH)_2$?

Use the Periodic Table to help you.

AO2 [2 marks]

Explaining neutralisation & energy changes

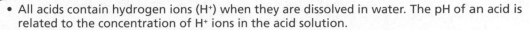

Explaining neutralisation

- When an acid reacts with an alkali, a salt and water are always made.

- All acids contain hydrogen ions (H^+) when they are dissolved in water. The pH of an acid is related to the concentration of H^+ ions in the acid solution.

- All alkalis contain hydroxide ions, OH^-, when they are dissolved in water.

- In neutralisation reactions the hydrogen ions and the hydroxide ions join up to form water molecules:

$$H^+(aq) + OH^-(aq) \longrightarrow H_2O(l)$$

- This is the **ionic equation** for all neutralisation reactions. The negative ion from the acid and the positive ion from the alkali are left in the solution to form the salt.

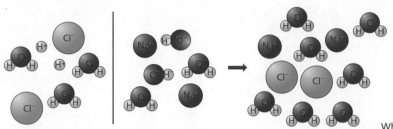

hydrochloric acid solution   sodium hydroxide solution   sodium chloride + water

When the ions in an acid and an alkali are mixed, water molecules are formed.

Remember!
Hydrochloric acid always forms chloride salts, sulfuric acid forms sulfate salts, and nitric acid forms nitrate salts.

- The positive ion from the alkali and the negative ion from the acid make the salt.

- The formula of the salt can be worked out by looking at the charges on the two ions.
  For example:

  sodium hydroxide + hydrochloric acid ⟶ sodium chloride + water

  Salt: sodium chloride. Positive ion: Na^+; negative ion: Cl^-;

  so formula: NaCl.

Higher

Energy changes

- **Exothermic** reactions give out heat energy. The temperature of the surroundings rises.

- **Endothermic** reactions take in heat energy. The temperature of the surroundings falls.

- An **energy level diagram** summarises the energy changes in a reaction.

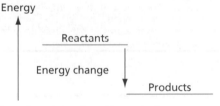

In exothermic reactions, the reactants give out energy, so the reactants are always at a higher energy value than the products.

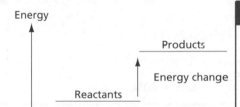

For endothermic reactions, the reactants take in energy and so are always at a lower energy level than the products.

EXAM TIP

Make sure that you can draw energy level diagrams for exothermic and endothermic reactions. Notice that the arrows to represent the energy change always go from the reactants to the products.

- Energy changes for large scale-reactions in industry need to be carefully controlled, because very extreme temperature changes could cause overheating, explosions or fires.

◉ Improve your grade

Energy changes

Foundation: Sam does some experiments. She does four different reactions in test tubes and takes a note of the temperature before mixing and 60 s after mixing. Sam's results are shown in the table.

Experiment	Temperature before mixing (°C)	Temperature 60 s after mixing (°C)
1	17	21
2	18	11
3	16	18

Which reaction is the most exothermic? Explain how you can tell.   *AO3* [3 marks]

Separating and purifying

The importance of purity

- A **pure** substance has nothing else mixed with it.

- In industry, pure substances need to be separated from impurities, such as left-over reactants or other products, before they are used. Some impurities may be harmful.

- **Filtration** can be used to separate a solid from a liquid or from a solution.

Crystallisation

- Crystallisation is used to purify impure solid crystals. The process has several steps:

  1 *Dissolving* – **dissolve** the product in a small amount of hot water (use only the minimum amount of water necessary to dissolve the product).

  2 *Filtering* – filter off any solid impurities that do not dissolve. The solution that comes through the filter is the **filtrate**.

  3 *Evaporating* – the filtrate starts to crystallise as some of the water **evaporates** off. Cool the filtrate while the product continues to **crystallise**.

  4 *Filtering* – filter off the crystals, leaving any soluble impurities in the solution.

  5 *Drying* – dry the crystals in a **dessicator** or oven.

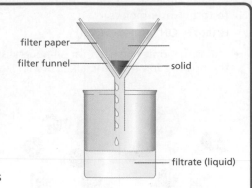

filter paper
filter funnel
solid
filtrate (liquid)

Filtering off solid impurities during crystallisation.

Percentage yield

- The percentage yield at the end of an experiment is worked out from the actual yield and the theoretical yield.

- The actual yield is the mass of product measured at the end of the experiment.

- The theoretical yield is the predicted yield. It is calculated from the amount of reactants used and the equation for the reaction.

- Percentage yield $= \dfrac{\text{actual yield}}{\text{theoretical yield}} \times 100\%$

- For example, an experiment reacts 2.4 g magnesium with hydrochloric acid. The actual yield of magnesium chloride is 5.7 g. What is the percentage yield?

Step 1: work out the theoretical yield:

Equation: $\quad Mg + 2HCl \longrightarrow MgCl_2 + H_2$

Relative masses: 24 $\qquad\qquad$ 95

Reacting masses: 2.4 g $\qquad$ 9.5 g

Theoretical yield of magnesium chloride = 9.5 g

Step 2: work out the percentage yield:

Percentage yield $= \dfrac{5.7}{9.5} \times 100\% = 60\%$

Remember!

The theoretical yield is worked out by looking at the equation for the reaction. The actual yield is what you measure at the end of your experiment. You need both values to work out the percentage yield.

EXAM TIP

Make sure you know how to work out theoretical yields from reacting amounts. At Higher tier, you will probably be asked to work out a theoretical yield first and then a percentage yield.

Improve your grade

Percentage yield

Higher: Ben prepares some copper sulfate crystals. He talks to Liz about his method.

Ben: 'First I added solid copper carbonate to sulfuric acid until the fizzing stopped.

Then I filtered off some unreacted copper carbonate.

Next I heated the filtrate to evaporate some of the water and left the solution to cool for a few minutes until some crystals formed.

Then I filtered off the crystals and weighed them. I made 1.4 g of crystals. I worked out that my theoretical yield is 1.6 g.'

Liz: 'Your percentage yield will be really inaccurate because you have missed some steps out of your method.'

Calculate Ben's percentage yield. Explain why his percentage yield is likely to be inaccurate. *AO2* [4 marks]

Rates of reaction

Measuring rates of reaction

- The **rate of reaction** is the amount of a product produced or the amount of reactant used up in a certain time. It is usually measured as the amount per second.

- Chemical engineers look for ways to control reactions. By speeding the reaction up they can make them more economical. They also need to ensure that reactions occur at a safe rate.

- In industry, chemical engineers aim to produce the most amount in the minimal time. They change conditions to make reactions faster to make the process as economical as possible, but they also consider the cost of the energy and safety.

- If the reaction makes a gas, the rate can be followed by:
  - measuring the volume of gas made at set times, e.g. every 30 seconds.
  - measuring the decrease in mass of the flask as the gas leaves the reaction.

- If the reaction makes a solid, the rate can be followed by measuring the time taken until you cannot see a cross underneath the flask or beaker.

- A colorimeter can be used to follow the rate of a colour change.

- Rate of reaction graphs show the change in the amount of reactant or product against time.

- The gradient of the curve at any point gives the rate of reaction.

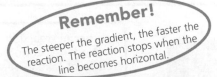
Remember!
The steeper the gradient, the faster the reaction. The reaction stops when the line becomes horizontal.

Changing rates of reactions

- For reactions to happen, particles must collide. The more collisions the faster the reaction.

- A **catalyst** is a substance that speeds up a chemical reaction but is not used up.

- Reactions are faster when:
  - the temperature of the reactants increases
  - the size of solid particles are smaller (this increases the surface area)
  - the concentration of reactants in solution increases (concentration is measured in grams per dm^3 of solution).

- To investigate the effect of changing one of these factors on the rate of the reaction, it is important that all other factors are kept constant.

The greater the concentration of reactants, the greater the collision frequency.

low concentration

- The rate of reaction increases when the frequency of collisions increases.

- Rate of reaction increases with a larger surface area of solids and with a higher concentration of solutions because the frequency of collisions increases.

- Chemical engineers control rates by controlling the factors that affect the rate, e.g. temperature, concentration, particle size or using a catalyst.

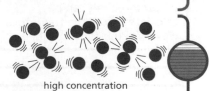

high concentration

Ideas about science

You should be able to:

- discuss how rates of reaction are controlled so that the benefits of faster reactions outweigh the costs

- discuss the costs to the environment of changing reaction conditions to increase the rate of industrial processes, e.g. the effects of using large amounts of energy to run a process at a high temperature

- discuss how using a catalyst makes industrial chemical reactions more sustainable.

Improve your grade

Changing rates of reactions

Foundation: Jack investigates the rate of the reaction between a large lump of calcium carbonate and dilute hydrochloric acid. He measures the volume of gas given off every 30 seconds.

The reaction takes place very slowly.

Suggest what changes Jack could make to his experiment to make the reaction faster. *AO1* [3 marks]

C6 Summary

The chemical industry uses large-scale chemical synthesis to make useful chemical compounds.

The pH of acids and alkalis can be tested using indicators or a pH meter.

Acids react with metals, metal oxides, metal hydroxides and metal carbonates to give salts and other products. These reactions can be summarised in equations.

Making chemicals, acids and alkalis

Relative formula masses can be worked out using relative atomic masses.

Reacting amounts and titrations

Balanced equations can be used to work out masses of reactants and products in reactions.

Titrations can be used to measure the exact amount of acids and alkalis that react together.

Acid and alkali reactions can be shown as an ionic equation between hydrogen ions and hydroxide ions.

$$H^+ + OH^- \rightarrow H_2O$$

Explaining neutralisation and energy changes

Exothermic reactions give out heat energy; endothermic reactions take in heat energy.

Energy changes in reactions can be shown on energy level diagrams.

Products of chemical reactions must be separated from impurities.

The percentage yield for preparing a salt can be calculated from the theoretical yield and the actual yield.

Separating and purifying

Solids can be separated from liquids and solutions by filtration. Crystals can be purified by crystallisation.

Rate of reaction depends on temperature, concentration of solutions and surface area of solids.

Rates of reaction

Rate of reaction increases when the frequency of particle collisions increases.

Catalysts increase the rates of reactions but are not used up.

The chemical industry

Bulk and fine chemicals

- **Bulk chemicals** are produced on a large scale in very large quantities.
- Some examples of bulk chemicals are ammonia, sulfuric acid, sodium hydroxide and phosphoric acid.
- **Fine chemicals** are produced on a small scale in much smaller quantities.
- Some examples of fine chemicals are drugs, food additives and fragrances.

Remember!

Bulk chemicals are made on a large scale, fine chemicals are made on a small scale.

Working in the chemical industry

- Chemists work in a range of different roles in the chemical industry.
- Chemists work to research and develop new chemical products and processes.

EXAM TIP

Some exam questions may ask you to identify which roles are done by chemists. Look for roles that involve using chemistry rather than making decisions about other aspects of work, such as finance or transport.

- Developing new products or processes involves an extensive programme of research and development.
- One example is the development of catalysts for new processes.

Government laws and the chemical industry

- Governments make laws and regulations to control different aspects of the chemical industry.

- The laws are designed to protect the safety of people and to reduce any impacts on the environment.
- These laws include regulations about the control of chemical processes in industry.
- There are also regulations about the safe storage and transport of chemicals.

When chemicals are transported their hazards must be displayed clearly. The letters and numbers tell the emergency services how to deal with spills.

Ideas about science

You should be able to:

- Write about how scientists limit the effects of the chemical industry on the environment by considering how chemical processes are run and the safe storage and transport of chemicals.

Improve your grade

Transporting chemicals

Foundation: By law, tankers containing petrol must be clearly labelled with information for the fire service. Give reasons why this is necessary.

AO3 [3 marks]

Green chemistry

Using chemical processes to make chemicals

- Chemical processes have a number of stages.
- The first stage is preparing the **feedstocks** (the chemicals the process uses).
- **Synthesis** happens when the chemicals react together to make the product.

- Sustainable processes should use resources that can be replaced and do not damage the environment.
  - Sustainable processes use renewable feedstocks.
  - Sustainable processes do not damage the environment at any stage of the process.

Products, by-products and sustainability

- At the end of the process, the products are separated from the rest of the mixture.
- **By-products** and waste need to be handled separately.
- The purity of the product is checked.

EXAM TIP

Be careful not to mix up 'renewable' and 'sustainable'. A renewable resource is not always sustainable. It might not be possible to make more of the resource quickly enough to meet the needs of a process.

- Sustainable processes consider these factors:
  - Using renewable feedstocks.
  - Finding other uses for by-products and disposal of waste.
  - Energy inputs and outputs and the use of renewable energy.
  - Reducing the environmental impact of each stage of the process.
  - Ensuring the health and safety of people.
  - Social and economic benefits for people.

Mining sulfur from an active volcano is a dangerous and unhealthy job.

Atom economy

- Reactions with high atom economy produce less waste.
- Reactions that use all of the atoms in the product have 100% atom economy.

- Atom economy is the percentage the mass of the atoms in the chemicals used in a reaction that end up in the product.

Ideas about science

You should be able to:

- Apply ideas about sustainability to chemical processes.

Improve your grade

Making zinc sulphate

Higher: The two equations show different ways of making zinc sulfate.

Equation 1: $Zn + H_2SO_4 \rightarrow ZnSO_4 + H_2$

Equation 2: $ZnCO_3 + H_2SO_4 \rightarrow ZnSO_4 + H_2O + CO_2$

Which equation has the higher atom economy? Explain your reasoning.

AO3 [3 marks]

Energy changes

Explaining endothermic and exothermic reactions

- **Exothermic** reactions give out heat energy to the surroundings.
- In exothermic reactions, reactants lose energy when they form products.
- **Endothermic** reactions take in heat energy from the surroundings.
- In endothermic reactions, reactants gain energy when they form products.

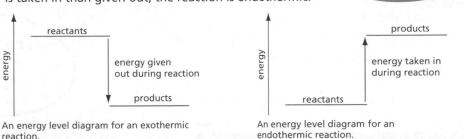

Remember!
For **exothermic** reactions, reactants are always **higher** than products. For **endothermic** reactions, reactants are always **lower** than products.

- Energy is taken in when bonds break.
- Energy is given out when bonds form.
- If more energy is given out than taken in, the reaction is exothermic.
- If more energy is taken in than given out, the reaction is endothermic.

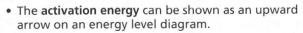

An energy level diagram for an exothermic reaction.

An energy level diagram for an endothermic reaction.

Energy level diagrams

- Energy changes can be shown using an **energy level diagram**.
- In exothermic reactions, reactants are higher than products on an energy level diagram.
- In endothermic reactions, reactants are lower than products on an energy level diagram.

- The **activation energy** can be shown as an upward arrow on an energy level diagram.
- The activation energy is the minimum amount of energy needed to start a reaction.
- The activation energy is used to break bonds to start a reaction.

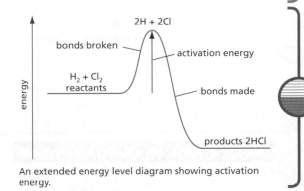

An extended energy level diagram showing activation energy.

- If the activation energy is high, very few molecules may have enough activation energy to start reacting.
- If the activation energy is low, many more molecules may have enough activation energy to start reacting.

Activation energy and rate

- Reactions with very high activation energies are usually slower than reactions with low activation energy.

- If an exothermic reaction has a large activation energy, it may not take place at room temperature.
- Activation energy can be supplied using a spark or heat to start the reaction.

Improve your grade

Energy changes and bonds

Higher: The reaction between hydrogen and chlorine is exothermic. The diagram shows molecules involved in the reaction.

H-H + Cl-Cl ⟶ H-Cl
                     H-Cl

Use ideas about energy changes and bonds to explain why this reaction is exothermic.   *AO2* [3 marks]

C7 Further chemistry   43

Catalysts and enzymes

Catalysts

- **Catalysts** increase the rate of reactions.
- Catalysts are not used up in reactions.
- Many industrial chemical processes use catalysts so that the products can be made faster.

- In industry, different catalysts are used for different reactions.
- A catalyst can work for many different reactions, but usually a particular process always uses a specific catalyst.

Enzymes

- **Enzymes** are biological catalysts.
- Enzymes are protein molecules that are found in living cells.
- Some industrial processes use enzyme catalysts.

- Enzymes usually only work as catalysts for one particular reaction.

Remember!
Enzymes are proteins. They do not work in extreme conditions, such as at high temperatures.

How do catalysts and enzymes work?

- Both catalysts and enzymes speed up reactions.

- Catalysts and enzymes increase the rate of a reaction by providing an alternative route for the reaction.
- A reaction that uses a catalyst has a lower **activation energy** than the same reaction without a catalyst.
- When the activation energy is lower, more reacting particles have enough energy to start reacting. This is why catalysts make reactions happen faster.

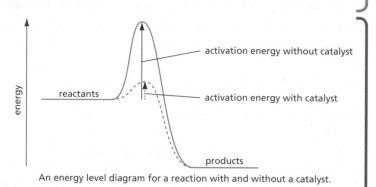

An energy level diagram for a reaction with and without a catalyst.

Restrictions on using enzymes

- Catalysts work across a much broader range of conditions than enzymes.

- Enzymes are very sensitive to reaction conditions so industrial processes that use enzymes can only work under a narrow range of conditions.

- Some reactions with high atom economy are too slow to be used in industry. Using a catalyst or enzyme in these reactions increases the rate. This means that reactions with high atom economy can be used in industry, which increases the sustainability of industrial processes.
- Using catalysts and enzymes also means that the reactions can happen at a lower temperature, saving energy and costs in industrial processes.
- Enzymes are **denatured** when their shape changes due to a change in conditions (for example at high temperatures or a change in pH).
- If the shape of an enzyme changes, it cannot work because it is the wrong shape to bind to reacting molecules.

Improve your grade

Catalysts and enzymes

Foundation: What are the similarities and differences between the way catalysts and enzymes work?

AO2 [4 marks]

Energy calculations

Bonds and chemical reactions

- During a chemical reaction, bonds are broken and formed.
- For a reaction to start, there must be enough energy available to break bonds.

- If the energy given out when bonds form is more than the energy taken in to break bonds then the overall reaction is exothermic.
- If the energy given out when bonds form is less than the energy taken in to break bonds then the overall reaction is endothermic.
- These energy changes can be shown on an energy level diagram.

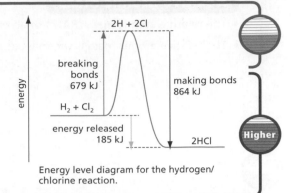

Energy level diagram for the hydrogen/chlorine reaction.

 Higher

Activation energies and bonds

- The minimum amount of energy needed to start a reaction is the **activation energy**.
- The activation energy provides the energy needed to break bonds.

- Different reactions have different activation energies because different bonds need to be broken.

Remember!
Energy is taken in when bonds break. Energy is given out when bonds form.

 Higher

Bond energies and energy changes

- Different types of bond have different bond energies.
- The energy change during a reaction can be worked out from the bond energies of all the bonds in the reactants and products.

- Step 1 is to work out the amount of energy given out when all of the bonds in the products form.
  - Make a list of every type of bond in the products.
  - Work out how many of each type of bond there are in the products.
  - To work out the total energy for each bond, multiply the number of each bond by the bond energy.
  - Add all of these bond energies together.
- Step 2 is to work out the amount of energy taken in to break all of the bonds in the reactants.
  - Make a list of every type of bond in the reactants.
  - Work out how many of each type of bond there are in the reactants.
  - To work out the total energy for each bond, multiply the number of each bond by the bond energy.
  - Add all of these bond energies together.
- Work out the overall energy change for an exothermic reaction using this formula.

$$\text{energy given out in an exothermic reaction} = \text{energy given out making bonds} - \text{energy taken in breaking bonds}$$

- The value of the energy change shows the energy given out in the reaction.

Higher

Improve your grade

Using bond energies

Higher: Fluorine reacts with hydrogen to form hydrogen fluoride.

$$H_2 + F_2 \longrightarrow 2HF$$

The bond energies for this reaction are shown in the table.

Bond	Energy (kJ/mol)
H-H	436
F-F	159
H-F	568

Calculate the energy change of reaction.

AO2 [3 marks]

Reacting masses

Relative atomic mass

- The **relative atomic mass** (RAM) of an element is shown on the Periodic Table.
- The relative atomic mass is the mass of an atom compared to the mass of carbon atom with a mass of 12.

relative
atomic mass

symbol

name

proton
number

32	35.5
S	**Cl**
sulfur	chlorine
16	17

The relative atomic mass of an element is the figure above the symbol.

Relative formula mass

- The **relative formula mass** (RFM) is the mass of a molecule or formula unit of a compound.
- Relative formula mass is also measured relative to the mass of a carbon atom with a mass of 12.
- The relative formula mass is worked out by adding together the relative atomic masses of every atom in the formula.

Chlorine molecules (Cl_2) are made up of two chlorine atoms (RAM 35.5).
The RFM is $2 \times 35.5 = 71$

$35.5 + 35.5 = 71$

Sodium sulfate has the formula Na_2SO_4
(RAMs Na = 23, S = 32, 0 = 16).
The RFM is $(2 \times 23) + 32 + (4 \times 16) = 142$

$23 + 23 + (32 + (4 \times 16)) = 142$

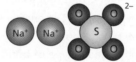

Masses of reactants and products

- The relative masses of the reactants and products in an equation can be used to work out the actual masses of reactants used and products formed.
- The first stage is to work out the relative formula masses of the reactants and products formed.

- The actual masses of reactants used and products formed are worked out from the equation in a series of stages.
  - Work out the relative atomic mass or relative formula mass of each reactant or product involved in the question.
  - Link the values together in a simple sentence.
  - Scale down to 1 g.
  - Scale up to the values in the question.

Higher

EXAM TIP

Make sure that you check the equation carefully – you may need to multiply the relative mass by two if there are two molecules or relative formula mass units in the equation.

Improve your grade

Calculating masses

Higher: Hydrogen can be used as a fuel.

$$2H_2 + O_2 \longrightarrow 2H_2O$$

What mass of oxygen is needed to completely burn 1 g of hydrogen?

AO2 [3 marks]

Alkanes

Structure of alkanes

- **Alkanes** are a family of hydrocarbons.
- **Hydrocarbons** are compounds that contain hydrogen and carbon only with no other elements.

Name	Molecular formula
methane	CH_4
ethane	C_2H_6
propane	C_3H_8
butane	C_4H_{10}

Some members of the alkane family.

Patterns in the properties of alkanes

- When alkanes (and all other hydrocarbons) burn in plenty of air, they make carbon dioxide and water.
- Alkanes do not dissolve in water and do not usually react with other reactants dissolved in water.

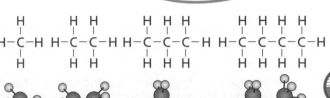

Remember!
Carbon forms 4 bonds – check any structures you draw to make sure that the carbon atoms have 4 bonds.

- Alkanes can be shown using 'ball and stick' diagrams to show the structures of their molecules.
- The bonds between the atoms in alkanes are all single covalent bonds.
- The C-C and C-H bonds in alkanes are difficult to break.
- The activation energy for reactions of alkanes is high because the bonds are difficult to break.
- This means that, although alkanes burn, they are generally unreactive and do not react with other reactants dissolved in water (aqueous solutions).

methane   ethane   propane   butane

Structural formulae show the bonds in the molecules and 'ball and stick' models show how the atoms are arranged in 3-D.

- When alkanes burn, energy must be taken in to break the bonds in the alkane and oxygen to start the reaction. Energy given out during the reaction to form the bonds in carbon dioxide and water is greater than the energy taken in. The reaction is exothermic and the burning continues.

methane   +   oxygen   →   carbon dioxide   +   water

Saturated and unsaturated

- Some hydrocarbons are saturated.
- **Saturated** molecules (like alkanes) contain only single bonds between two carbon atoms.
- **Unsaturated** molecules (e.g. ethene) contain one or more double bonds between two carbon atoms.
- Alkanes have the general formula $C_nH_{(2n+2)}$.
- Alkenes have the general formula $C_nH_{(2n)}$.

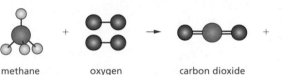
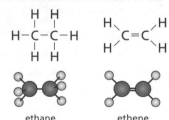

ethane   ethene

Ethane is a saturated alkane and ethene is unsaturated.

Improve your grade

Formulae of alkanes

Higher: The table shows information about the formulae of some alkanes.

Alkane	Number of carbon atoms	Formula
Ethane	2	C_2H_6
Propane	3	C_3H_8
Butane	4	C_4H_{10}

The next alkane in the series is called pentane.

What is the formula of a molecule of pentane? Explain how you worked out your answer.   *AO2* [3 marks]

Alcohols

The alcohol family

- The **alcohols** are a family of compounds.
- The alcohol with the smallest molecule is methanol, CH_3OH.
- Methanol is a toxic liquid that is used as a fuel and solvent.
- The type of alcohol in alcoholic drinks is ethanol, C_2H_5OH.
- Ethanol is also a liquid and is also used as a fuel and a solvent.

- Alcohols contain the **functional group** –OH.
- The oxygen and hydrogen atoms are bonded together by a covalent bond and also bond to the rest of the alcohol molecule by a covalent bond.

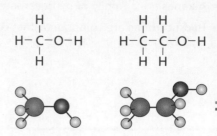

methanol   ethanol
Structural formulae and ball and stick models of simple molecules.

Structure and properties of alcohols

- Alcohols burn in air to make carbon dioxide and water.
  e.g. ethanol + oxygen ⟶ carbon dioxide + water

- The –OH group means that alcohols have different properties to alkanes.
  – Alkanes are unreactive and are not soluble in water.
  – All alcohols are soluble in water.
  – Each alcohol has a higher melting and boiling point compared to an alkane with the same number of carbon atoms.
- Alcohols are flammable because they contain a hydrocarbon chain of carbon and hydrogen atoms.

> **EXAM TIP**
>
> Make sure that you can compare the properties of alcohols to alkanes.

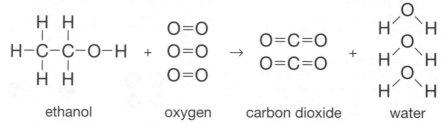

ethanol   oxygen   carbon dioxide   water

C-C, C-H and C-O bonds in the ethanol break when the molecule burns.

Reaction between alcohols and sodium

- Ethanol reacts with sodium.
- The piece of sodium fizzes as the hydrogen gas is made.
- Ethanol reacts with sodium to give hydrogen and sodium ethoxide.
  sodium + ethanol ⟶ sodium ethoxide + hydrogen
  $2Na(s) + 2C_2H_5OH(l) \longrightarrow 2C_2H_5ONa(s) + H_2(g)$

- This is similar to the reaction between water and sodium.
- The reaction between sodium and ethanol is slower than between sodium and water.
- Ethanol reacts with sodium because it has a –OH group.
- Alkanes do not react with sodium because they have no –OH group.

> **Remember!**
>
> Sodium reacts with water to make hydrogen and sodium hydroxide.

⦿ Improve your grade

Reacting sodium with propanol

Higher: Propanol is an alcohol.

A piece of sodium is dropped into propanol and into water.

Predict how the reaction of sodium with propanol will differ from the reaction of sodium with water.

AO3 [2 marks]

Fermentation and distillation

Fermentation

- **Fermentation** is a process that is used to make alcoholic drinks.
- Ethanol is the alcohol in alcoholic drinks.
- During fermentation yeast turns sugar into ethanol.
- Yeast is a microorganism.
- The yeast uses the sugar as a food to make energy.
- Ethanol and carbon dioxide are formed as waste products.

- Fermentation can be represented by an equation.

    sugar → ethanol + carbon dioxide

- Air must be kept out during fermentation because ethanol is made when yeast respires anaerobically (without oxygen).

Why fermentation stops

- Fermentation can only be used to make a dilute solution of ethanol.
- Fermentation stops when the ethanol becomes more concentrated.

- Yeast produces enzymes that cause fermentation.
- Higher concentrations of ethanol kill the yeast.

- Ethanol is toxic at higher concentrations.
- The ethanol produced by fermentation is toxic to yeast. When the concentration rises, the yeast dies.

Fermentation and enzymes

- Yeast is needed to produce enzymes for fermentation.

- Enzymes only work in a narrow temperature range. If the reaction mixture is too low or too high, fermentation stops.
- Yeast is also affected by pH. The enzymes work best at about pH5. If the pH is too high or too low, fermentation stops.

> **Remember!**
> Enzymes are biological catalysts. They work under only a narrow range of temperature and pH ('optimum conditions').

- If conditions of temperature or pH are outside the optimum range for the enzymes, fermentation will slow down or stop. This can happen even at low concentrations of ethanol.

Distillation

- **Distillation** is a process that is used to make a dilute solution of ethanol more concentrated.
- Distillation involves several stages.
    - The dilute solution of ethanol is heated to just above the boiling point of ethanol (78 °C).
    - The ethanol boils off and leaves most of the water behind.
    - The ethanol vapour is cooled and condensed to form a liquid.
    - The liquid contains a more concentrated solution of ethanol, called a 'spirit'.
- Distillation is used to make ethanol for fuel and spirit drinks such as whisky and brandy.

- Distillation requires large amounts of energy to evaporate the ethanol.
- Using energy has an impact on the cost of the process and the environment.

Improve your grade

Making wine

Higher: Joe has a hobby making wine. He finds that one batch of wine has stopped fermenting. Suggest some reasons why the wine may have stopped fermenting. *AO1* [4 marks]

Alternatives to fermentation

Making ethanol from plants

- Fermentation uses sugar from plants grown as crops, which contain only small amounts of sugar.
- Large amounts of unused plant material and carbon dioxide gas are waste products of the process.
- Other ways of making ethanol may be more sustainable.
- Waste plant material from agriculture can be used as **biomass**.
- Scientists are developing new processes to convert biomass into ethanol.
- Some processes use genetically modified bacteria.

> **Remember!**
> Not all processes that use renewable resources are sustainable. Some processes can still cause environmental harm over time.

- Biomass can be converted into ethanol using genetically modified *E.coli* bacteria.
  - Plant material from a range of sources can be used.
  - The biomass is dissolved in acids and solvents.
  - The *E.coli* bacteria use cellulose in the plant material as food and produce ethanol.
  - The process works at just above normal temperatures.
- Producing ethanol using biomass does not yet work well enough to be used on a large scale.

- Making ethanol from biomass may be sustainable in the long term.
  - Plant material is renewable, and the process can use waste plant material that has no other uses.
  - The acids and solvents used to dissolve the biomass may be harmful to people or the environment.
  - The conditions for the process use little energy.
  - Eventually it may be possible to convert about 90% of the plant cellulose into ethanol, giving high atom economy.
  - The processes are not yet fully developed.

Making ethanol from crude oil

- Crude oil can be used to make ethanol.
  - Ethane is an alkane in crude oil.
  - Ethane can be converted into ethene (a molecule with a double bond).
  - Ethene reacts with steam to make ethanol.
  - A catalyst is needed for the reaction.
  - At the end of the process, distillation is used to separate and purify the ethanol.

- The process used to make ethanol from ethane involves two stages:
  - In stage 1, ethane reacts to make ethene and hydrogen.
  - In stage 2, ethane from stage 1 reacts with steam to make ethanol.
  - this reaction happens at a high temperature (300 °C) and uses a catalyst (phosphoric acid).

Stage 1

Stage 2

- Making ethanol from crude oil is not sustainable in the long term.
- Making ethanol from crude oil has both advantages and disadvantages.
  - Crude oil is not renewable.
  - The process runs at a high temperature and so needs large amounts of energy.
  - The atom economy of the reactions have high atom economy because only hydrogen is made.
  - Hydrogen is an excellent by-product because it has many uses.
  - The process is much faster than using fermentation.

Improve your grade

Making ethanol

Higher: Describe the advantages and disadvantages of making ethanol from crude oil.

AO1 [3 marks]

Carboxylic acids

What are carboxylic acids?

- **Carboxylic acids** are a family of organic compounds.
- Carboxylic acids have a sharp, unpleasant taste.
- Many carboxylic acids have unpleasant smells.
- Carboxylic acids cause the smell of sweaty socks and the taste of rancid butter.

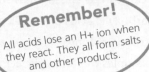

Remember!
All acids lose an H+ ion when they react. They all form salts and other products.

- Carboxylic acids have a hydrocarbon chain (like in alkanes) and a carboxylic acid functional group –COOH.
- The –COOH functional group is the part of the molecule that gives the acidic properties of carboxylic acids.

- The –COOH group can lose an H$^+$ ion to form a negative ion.
- The negative ion formed from the carboxylic acid contains the hydrocarbon chain attached to –COO$^-$.

Methanoic and ethanoic acid

- **Methanoic acid** (HCOOH) and **ethanoic acid** (CH$_3$COOH) are both carboxylic acids.
- Some insect and plant stings contain methanoic acid.
- Vinegar contains ethanoic acid.

- The formula of methanoic acid is HCOOH and the formula of ethanoic acid is CH$_3$COOH.

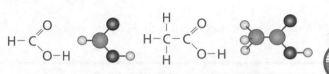

methanoic acid   ethanoic acid

Structural formulae and ball and stick diagrams of methanoic acid and ethanoic acid.

Properties of carboxylic acids

- All carboxylic acids are typical acids. They have a low pH and they react with metals, carbonates and alkalis.

- Carboxylic acids react with metals, alkalis and carbonates to give salts and other products.
  carboxylic acid + metal ⟶ salt + hydrogen
  e.g. ethanoic acid + magnesium ⟶ magnesium ethanoate + hydrogen
  carboxylic acid + alkali ⟶ salt + water
  e.g. ethanoic acid + sodium hydroxide ⟶ sodium ethanoate + water
  $$CH_3COOH \ + \ NaOH \ \longrightarrow \ CH_3COONa \ + \ H_2O$$
  carboxylic acid + metal carbonate ⟶ salt + water
  e.g. ethanoic acid + sodium carbonate ⟶ sodium ethanoate + carbon dioxide + water
  $$2CH_3COOH \ + \ Na_2CO_3 \ \longrightarrow \ 2CH_3COONa \ + \ CO_2 \ + \ H_2O$$
- The names of salts made by carboxylic acids all have the ending '-oate'.

- The names of salts made from carboxylic acids all have the ending '-oate', e.g. sodium methanoate is a salt with the formula HCOONa. It contains a negative methanoate ion HCOO$^-$ and a positive sodium ion Na$^+$.

Improve your grade

Reactions of carboxylic acids

Higher: Complete the table to show the products of some reactions of carboxylic acids. *AO2 [3 marks]*

Reactants	Products
methanoic acid + magnesium	
ethanoic acid + potassium hydroxide	
propanoic acid + sodium carbonate	

Weak acids

Strong and weak acids

- Acids can be classified as strong or weak.

- Carboxylic acids are **weak acids**.

- Strong acids are corrosive at all concentrations. Weak acids are only corrosive at high concentrations.

- Examples of carboxylic acids include methanoic acid (HCOOH) and ethanoic acid (CH_3COOH).

- Ethanoic acid is used to make vinegar. It is safe to use in food because vinegar is a dilute solution and also because ethanoic acid is a weak acid.

Vinegar is made from ethanoic acid.

- Examples of strong acids include hydrochloric acid, sulfuric acid and nitric acid.

pH Strong and weak acids

- Strong acids have a very low **pH**.

- Weak acids such as carboxylic acids have a higher pH than strong acids (closer to neutral pH7).

- Strong acids have a lower pH than weak acids because they produce a higher concentration of H^+ ions when they dissolve in water (more H^+ ions per cm^3).

- All acids produce H^+ ions when they react. Acid reactions almost always happen when the acids are dissolved in water.

- Strong acids ionise to make H^+ ions more easily than weak acids.
  - Strong acids such as hydrochloric acid (HCl) ionise completely into ions when they dissolve in water (to make H^+ and Cl^-).

- Most of the molecules in weak acids do not ionise when they dissolve in water.
  - When ethanoic acid (CH_3COOH) dissolves in water, most of the molecules stay as complete molecules. Only a very few molecules ionise to make H^+ and CH_3COO^- ions.

Reactions of strong and weak acids

- Weak acids are less reactive than strong acids.

- Weak acids react more slowly than strong acids.

- The rate of reaction of strong and weak acids are different with metals and metal carbonates.
  - Carboxylic acids react more slowly than strong acids with metals such as magnesium.
  - Carboxylic acids react more slowly with metal carbonates such as magnesium carbonate.
  - Both of these reactions give off a gas. Carboxylic acids fizz more slowly than strong acids because the gas is made more slowly.

> ### EXAM TIP
> Common questions ask you to compare the properties of a carboxylic acid with one of the strong acids. Make sure you know what products acids make when they react with metals and carbonates and remember to mention rate and pH too.

Improve your grade

Reactions of strong and weak acids

Foundation: Describe how you use magnesium metal to tell the difference between a strong and a weak acid. Include what observations you would make. *AO2* [3 marks]

Esters

Using esters

- Esters have fruity smells and flavours.
- Natural esters give fruits and flowers their distinctive tastes or smells.
- Esters have low boiling points and evaporate quickly at room temperature so that their smells travel easily.

- Esters are used as **solvents**.
- Ethyl ethanoate is an example of an ester. It is used as nail varnish remover because it acts as a solvent.
- Some paints and inks contain solvents that are esters.
- Esters are used as fruity food flavourings and perfumes.
- Esters are used as plasticisers. They are added to plastics to make them more flexible.

Making esters

- Esters are made when a carboxylic acid reacts with an alcohol.

- Synthetic esters can be made.
- Synthetic esters are often cheaper to produce than natural flavourings and perfumes.

Stages used to make esters

- An ester is made when a carboxylic acid reacts with an alcohol.
- Different esters are made by reacting together different carboxylic acids with different alcohols.
- Ethyl ethanoate is made by reacting ethanoic acid with ethanol.

ethanoic acid	+	ethanol	$\longrightarrow$	ethyl ethanoate	+	water
CH_3COOH	+	C_2H_5OH	$\longrightarrow$	$CH_3COOC_2H_5$	+	H_2O

- Ethyl ethanoate is made in very large quantities for using as a solvent in paints, glues, dyes, coatings and nail varnish remover.
- The reaction is very slow so the conditions need to be adapted to make the reaction faster.
  - A catalyst is used (usually concentrated sulfuric or phosphoric acid).
  - The mixture is heated.
- The techniques used to make a liquid ester, such as ester ethanoate, involve several stages.
  - Heating under **reflux**.
  - Distillation.
  - Purification, using a **tap funnel**.
  - Drying.

> **EXAM TIP**
>
> Make sure you can recognise an ester. Look for the 'COO' in the middle of the hydrocarbon chain.

> **EXAM TIP**
>
> Make sure you know each stage involved, what happens at each stage, and why they are important.

Higher

methyl methanoate   methyl ethanoate
Examples of esters.

Improve your grade

Nail varnish remover

Foundation: Ethyl ethanoate is used to make some types of nail varnish remover. Ethyl ethanoate is an ester. Explain the advantages of using ethyl ethanoate as a nail varnish remover. *AO2* [3 marks]

Making esters

Reflux and distillation

- Ethanoic acid and ethanol react together by heating under reflux.

- Heating under reflux is used to heat substances with low boiling points.

- During heating under reflux, a vertical condenser is attached to a flask. The contents of the flask evaporate when they are heated but then cool down and turn back to liquids in the condenser. They drip back down into the flask and are not lost.

- At the end of the reaction the ethyl ethanoate needs to be separated from the mixture. This is done using distillation.

- During distillation, the mixture is heated again, but this time the condenser is positioned at an angle.

- The ethyl ethanoate evaporates, cools and condenses in the condenser and can be collected as it drips from the condenser.

- Most of the mixture is left behind in the flask.

- **Higher** The ethyl ethanoate still contains impurities. It contains unreacted ethanoic acid and ethanol as well as water.

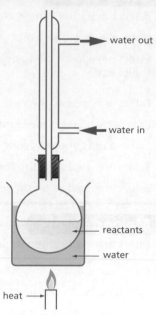

Ethanoic acid and ethanol being heated under reflux.

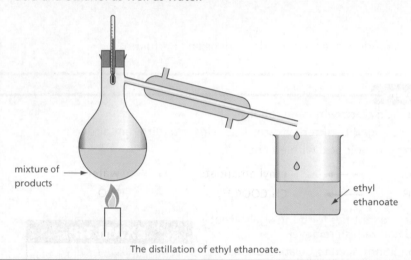

The distillation of ethyl ethanoate.

Purification in a tap funnel

- The ethyl ethanoate is purified in a tap funnel. Separate layers can be run out of a tap funnel.

- Ethanoic acid impurities are removed by adding sodium carbonate dissolved in water.

  - Ethanoic acid and sodium carbonate react to make a salt, carbon dioxide and water.

  - The carbon dioxide is a gas and can be released from the top of the tap funnel.

  - The ethyl ethanoate and water form separate layers.

  - The salt dissolves in the water and can be run off.

- **Higher** Ethanol is removed by adding calcium chloride dissolved in water. The ethanol dissolves in the calcium chloride and water layer and can be run off.

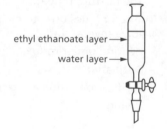

Drying

- **Higher** Solid anhydrous calcium chloride is mixed with the ethyl ethanoate. This dries the ester by removing all of the water. The waste solid can be filtered off.

- The final ethyl ethanoate may be distilled again to ensure it is pure. Only distillate that boils at the exact boiling point of ethyl ethanoate (77 °C) is collected.

Improve your grade

Making ethyl ethanoate

Higher: Ethyl ethanoate is made from ethanoic acid and ethanol. Sulfuric acid is added to the mixture before it is heated. Explain why this is necessary.　　　　　　　　　　*AO2* [3 marks]

Fats and oils

Fats and oils in living things

- Fats are solids at room temperature.
- Oils are liquids at room temperature.
- Living things use fats and oils for energy.
- Fats usually come from animals.
- Oils usually come from plants.
- Examples of oils from plants include oils used in food, such as olive oil and sunflower oil.

Remember!
All esters are made when a carboxylic acid reacts with an alcohol. Water molecules are also made in the reaction.

- Fats and oils are made from **fatty acids** and glycerol.
- All fats and oils are based on the glycerol molecule, but different fats and oils have different structures because the fatty acids are different.

Structure of fats and oils

- Fats and oils are esters.
- Fatty acids are carboxylic acids with a very long hydrocarbon chain.
- **Glycerol** is a tri-alcohol.
- Glycerol has three –OH groups.
- Fats and oils consists of three fatty acids which link by ester groups to the three –OH groups on the glycerol molecule.

carbon chains of various lengths

The molecular structure of a fat or oil.

- Most animal fats contain fatty acids with saturated hydrocarbon chains.
- All of the carbon atoms in a saturated hydrocarbon chain are joined by single carbon-carbon bonds.
- Most vegetable oils contain fatty acids with unsaturated hydrocarbon chains.
- Some of the carbon atoms in an unsaturated hydrocarbon chain are joined by double carbon-carbon bonds.
- Bromine water can be used as a test for unsaturation.
- An unsaturated compound turns bromine water from orange-brown to colourless.
- Saturated compounds do not react with bromine water.
- The bromine water test can also be used to tell the difference between an alkane and an alkene.

Testing oils for unsaturation. On the left an oil floats above bromine solution. After shaking, the solution turns colourless, indicating that the oil is unsaturated, as shown on the right.

Improve your grade

Esters and oils

Higher: Nail varnish remover contains ethyl ethanoate. Face cream contains oils to soften the skin.

Give one similarity and one difference between the structure of a molecule of ethyl ethanoate and a molecule of oil.

AO2 [3 marks]

Reversible reactions

Reversible reactions and equilibrium

- Most chemical reactions only work in the forwards direction. Products cannot react to form reactants again.

- **Reversible** reactions can happen in both the forwards and backwards direction.

- Reactants react to form products. Products can also react to form reactants again in a backwards reaction.

- The symbol used in an equation for a reversible reaction is $\rightleftharpoons$.

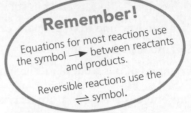

> **Remember!**
> Equations for most reactions use the symbol $\longrightarrow$ between reactants and products.
> Reversible reactions use the $\rightleftharpoons$ symbol.

ammonium chloride	$\rightleftharpoons$	ammonia	+	hydrogen chloride
$NH_4Cl(s)$	$\rightleftharpoons$	$NH_3(g)$	+	$HCl(g)$

- Reversible reactions can reach **equilibrium**.

- The $\rightleftharpoons$ symbol means that the reaction is reversible and can reach equilibrium.

What happens at equilibrium?

- Equilibrium can only be reached in a closed container.

- If the reactants or products escape, the reaction will not reach equilibrium. The closed container makes sure that the reactants and products cannot escape.

- At equilibrium, the amounts of the reactants and products stay the same (as long as the conditions stay the same).

- At equilibrium, the rate of the forwards reaction is the same as the rate of the backwards reaction.

- Equilibrium is 'dynamic' because reactions continue to happen, even at equilibrium.

- At equilibrium:
  - The forward reaction continues to happen.
  - The backwards reaction continues to happen.
  - The forwards and backwards reactions are happening at the same rate.

Higher

- At equilibrium the reactants are being used up and formed at the same rate.

- At equilibrium the products are being formed and used up at the same rate.

- There appears to be no change (as long as the conditions are kept the same).

- The container must be kept closed so that no reactants or products escape.

Reversible reactions

- Plants need nitrogen for growth.
  - The air contains a large amount of nitrogen gas. Nitrogen gas is an element.
  - Plants cannot use nitrogen gas from the air.
  - Plants use nitrogen in compounds. Plants take in nitrogen compounds called nitrates.
  - Forming nitrate compounds from nitrogen is called **fixing** nitrogen.
  - Nitrate compounds are formed naturally, but there are not enough natural nitrates available for growing large amounts of crops.
  - Farmers add fertiliser to soil to provide nitrates for crops.

- Synthetic fertilisers are used to provide crops with nitrogen for growth.

- Synthetic fertilisers replace natural nitrates in the soil but must be made using an industrial process and can cause environmental problems when they are used.

Improve your grade

Making sulfur trioxide

Higher: The equation shows the reaction between sulfur dioxide and oxygen to form sulfur trioxide. This reaction is important in the manufacture of sulfuric acid.

$$2SO_2 + O_2 \rightleftharpoons 2SO_3$$

Explain why this reaction cannot produce 100% yield of sulfur trioxide.   *AO2* [3 marks]

The Haber process

The Haber process

- Synthetic fertiliser is made from ammonia.

- Ammonia is made by the **Haber process**.
  - The Haber process is named after Fritz Haber. He invented the process.
  - The feedstocks for the process are hydrogen and nitrogen gas.
  - In the process nitrogen gas reacts with hydrogen gas.
  - Nitrogen gas comes from the air.
  - Hydrogen is made by reacting natural gas (methane) with steam.

    nitrogen + hydrogen $\rightleftharpoons$ ammonia

    $N_2 + 3H_2 \rightleftharpoons 2NH_3$

- The $\rightleftharpoons$ symbol shows that the reaction between nitrogen and hydrogen is reversible.

Choosing the conditions

- Choosing conditions for the industrial process involves considering both the rate of the reaction and the yield of ammonia.

- The rate of reaction for the Haber process is increased by:
  - Using a catalyst.
  - Raising the temperature.
  - Increasing the pressure (increasing pressure increases rate of reaction for gases).

- The reaction uses an iron catalyst at a temperature of 450 °C and 200 atm pressure.

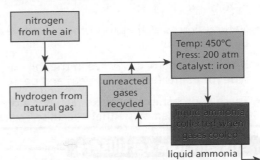

The Haber process

- The conditions for the process also affect the yield of ammonia (the amount of ammonia formed as a product).
  - The reaction is exothermic. Increasing the temperature decreases the yield (causing less ammonia to be made).
  - This is because a higher temperature increases the rate of the backwards reaction.
  - Increasing the pressure increases both the rate of reaction and the yield.

- The gases do not spend long enough in the reaction vessel to reach equilibrium.

- Only about 20% of the nitrogen and hydrogen react together to make ammonia.

- The ammonia is removed from the mixture that leaves the reaction vessel and the nitrogen and hydrogen are recycled so that no reactants are wasted.

- The conditions chosen are a compromise taking into account the rate of reaction, the yield and the cost of running the process.

- Choosing the temperature…
  - A higher temperature increases the rate of reaction (if the temperature is too low the reaction is too slow).
  - But a higher temperature decreases the yield of ammonia.
  - A higher temperature is more expensive (because of the cost of fuel).
  - A 'compromise' temperature of 450 °C gives a reasonable yield and rate and is not too expensive to maintain.

- Choosing the pressure…
  - A higher pressure increases both the rate of reaction and the yield.
  - A very high pressure means that the equipment is more likely to leak.
  - A very high pressure is more expensive to maintain.
  - A 'compromise' pressure of about 200 atm is used.

Improve your grade

Saving waste

Higher: In the Haber process only about 20% of the hydrogen and nitrogen react in the reaction vessel to make ammonia. Explain what is done to make sure that this does not result in a waste of feedstocks.

AO2 [3 marks]

Alternatives to Haber

Organic and synthetic fertilisers

- Organic fertilisers are made from waste material from animals and plants.
- Synthetic fertilisers are made from feedstocks in industrial processes.
- Organic fertilisers are generally much less soluble in water than synthetic fertilisers.
- Soluble fertilisers dissolve in rainwater and can get washed into rivers and the sea.
  - Fertiliser causes algae in the water to grow.
  - The algae dies and bacteria that live on the dead algae multiply, taking oxygen from the water.
  - Fish and other organisms in the water die from lack of oxygen.
  - Using too much fertiliser or putting fertiliser on the fields when crops are not growing makes the problem worse.

- Scientists are studying how enzymes from bacteria fix nitrogen directly from the air to try to copy natural processes to make fertiliser. They are trying to make new fertilisers that are similar to natural organic fertilisers.
- Natural processes that fix nitrogen involve enzymes that are made by bacteria.
- Nitrogen-fixing bacteria live in the soil and in the roots of some types of plants (e.g. clover, peas and beans).
- These approaches may make fertiliser more economically without using as much energy and also solve the problems caused by synthetic fertilisers when they are washed into rivers and the sea.

> **Remember!**
> Fixing nitrogen means taking nitrogen gas from the air and reacting it to make nitrogen compounds that can then be used to make fertiliser.

Sustainable fertilisers

- Scientists are using their understanding of how bacteria and their enzymes fix nitrogen to develop new approaches to fertilising crops.
  - Copies of natural enzymes are being developed to make fertilisers economically in processes that work at room temperature.
  - Genetically modified crops are being developed that can make their own enzymes to fix nitrogen.
  - Genetically modified crops are being developed that have bacteria living in their roots to fix nitrogen.

- A process is sustainable if it can meet the demands of people without causing lasting harm to the environment.
- Processes that rely on the use of bacteria and enzymes may be more sustainable in the long run.
  - These processes do not use as much energy.
  - They do not depend on non-renewable resources.
  - However, these process are still experimental and do not work well enough for large-scale use yet.
- Making fertiliser by the Haber process is not sustainable in the long run.
  - Natural gas (methane) is used as a raw material for making hydrogen. This is a non-renewable resource that will run out.
  - The process uses large amounts of energy for the high temperatures and pressures needed for the reaction.
  - Energy is also used for the large-scale transport of raw materials and products.
  - The process uses large amounts of water.
  - The process produces large amounts of carbon dioxide due to high energy use.

Ideas about science

You should be able to:

Use information given in a question to discuss why one way of making fertiliser is more sustainable than another.

Improve your grade

Using fertilisers

Foundation: Give one advantage and one disadvantage of using synthetic fertilisers rather than organic fertilisers to grow food crops.

AO1 [2 marks]

Analysis

Qualitative and quantitative analysis

- Chemical analysis involves using practical techniques to test a sample to collect information and data about the chemicals the sample contains.

- **Qualitative data** gives information to identify which substances the sample contains.

- **Quantitative data** gives numerical data about the amounts of substances the sample contains.

- Measurements on each sample are repeated. This is important because outliers or errors can be identified before the data is processed.

- Each stage of analysis procedures are standardised so that they are the same every time. This is so that measurements can be compared.

- Samples of a material are chosen so that they represent the bulk of the material under test. This is because in practice it is not possible to test every part of a material.

- Samples are prepared carefully for analysis. The amount of the sample is measured and great care is taken to avoid contamination.

Collecting and storing samples

- Care is taken by the person collecting the samples to make sure that they do not become contaminated with impurities. For example:

  - Wearing gloves and/or a mask.

  - Making sure that equipment used in collecting samples is very clean.

  - Making sure the container used to hold the sample is very clean and that the material used to make the container does not contain any chemicals that might get into the sample (e.g. acids in paper or plasticisers in plastics).

The person collecting samples has to be very careful.

EXAM TIP

You may be asked to describe what procedures you would use when collecting samples. You need to consider where you would take your samples, how many samples you would take, when to take the samples and how you would avoid contamination.

Storage and preparation

- Samples need to be carefully stored so that they do not deteriorate before they are tested.

- The method of storing samples depends on the type of the sample.

  - Samples that contain bacteria may be kept refrigerated to stop bacteria from multiplying.

  - Some samples may be stored in a freezer, but this is not suitable for all samples as freezing may cause damage (e.g. to plant and animal tissues if ice crystals form in cells).

  - Samples may be stored under very dry conditions.

  - Some are stored under an inert gas (such as argon or nitrogen) to stop oxidation by reaction with oxygen in the air.

⊙ Improve your grade

Sampling tablets

Foundation: Carol works for a factory that makes medicines in batches. The medicines are sold in bottles as tablets. The factory continues production for 24 hours every day. Carol's job is to check the contents of the tablets before they leave the factory.

How should Carol choose tablets to check to make sure that the checking is as thorough as possible?

AO2 [3 marks]

Principles of chromatography

Stationary phase and mobile phase

- Chromatography is used as a method of analysis that can be used for some mixtures.
- All types of chromatography involve a **stationary phase** and a **mobile phase**.
- The stationary phase does not move, e.g. the paper in paper chromatography.
- The mobile phase moves and carries the substances in the mixture along. The substances spread out as they move.
- Some types of chromatography separate substances that are dissolved to make a solution.
  - Water makes a good solvent because many substances dissolve in water.
  - An **aqueous** solution contains substances dissolved in water.
  - A **non-aqueous** solution contains substances dissolved in other solvents (e.g. ethanol or cyclohexane).
- Using different solvents in chromatography gives a different pattern because different solvents spread the substances out differently as the solvent moves.

- The way the substances spread out depends on how soluble each substance is in the solvent. More soluble substances will usually spread further.

Chromatography and solubility

- The substance that is dissolved in a solution is called the solute.
- A solute may be more soluble in some solvents than in others.

- Some solvents do not mix. They are **immiscible**. They form separate layers.
- A solute added to the solvents will dissolve in both.
- How much solute dissolves in each solvent depends on its solubility in each solvent.
- More solute will dissolve in one layer of solvent than in another. Therefore the solute has a different concentration in each layer.
- The solute is **distributed** between the two solvents.
- The movement of the molecules of the solute between the two solvents is reversible and reaches equilibrium.

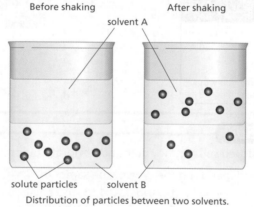

Before shaking   After shaking

solvent A

solute particles   solvent B

Distribution of particles between two solvents.

- Chromatography works because the substances in the mixture are distributed between the mobile phase and the stationary phase.
- The substance that is most soluble in the mobile phase will dissolve more and be carried along the most when the mobile phase moves. This means it will travel the furthest.
- The substance that is least soluble in the mobile phase will stay attached to the stationary phase and will move the least.
- Substances separate out depending on how they are distributed between the mobile phase and the stationary phase.

> **Remember!**
> In an equilibrium reaction the amount of substances stay the same because the rate of reaction in both directions is the same.

Improve your grade

Removing grease

Foundation: Dan has grease marks on his jeans. He washes them in a washing machine but the marks do not wash out. He uses a stain remover that contains a non-aqueous solvent to remove the marks. Explain why the stain remover removes the marks but washing does not. *AO2* [3 marks]

Paper and thin-layer chromatography

Method for chromatography

- In paper chromatography, the stationary phase (the phase that does not move) is a piece of chromatography paper.

- In thin-layer chromatography (TLC) the stationary phase is a plate of plastic or glass coated with a thin layer of solid (e.g. silica or alumina).

- A spot of sample is put on a pencil line high enough to be above the layer of the solvent.

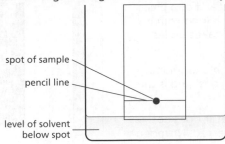

spot of sample

pencil line

level of solvent
below spot

An example of paper chromatography.

- The paper or plate is placed in the solvent. If a non-aqueous solvent is used, a lid needs to be used to stop the solvent evaporating.

- The solvent rises slowly upwards, separating the substances in the spot.

- The paper or plate should be taken out of the solvent before the solvent reaches the top.

- The distance moved by the solvent is marked.

- Coloured substances, such as inks, separate out to give different spots for each substance in the mixture.

- **Reference materials** are used to identify the spots. This involves spotting known substances at the same level as the spot of the unknown mixture at the start of the chromatogram.

Chromatography results

- The distances travelled by the reference material spots can be matched to the unknown spots to identify the unknown substances in the mixture.

- Some substances are not coloured. A **locating agent** needs to be used so that the spots can be seen.

- Chemicals (e.g. iodine or ninhydrin) or UV light can be used as locating agents.

- R_f values are calculated from the distance moved by a spot.

$$R_f \text{ value} = \frac{\text{distance moved by sample}}{\text{distance moved by solvent}}$$

- R_f values can be used to identify the substances in a mixture by comparing the R_f values of unknown substances with R_f values for known substances.

- R_f values are always the same for a particular substance as long as the same solvent is used at the same temperature.

- R_f values change if a different solvent is used or if the temperature is changed. This is because the solubility of the substance in the solvent will be different.

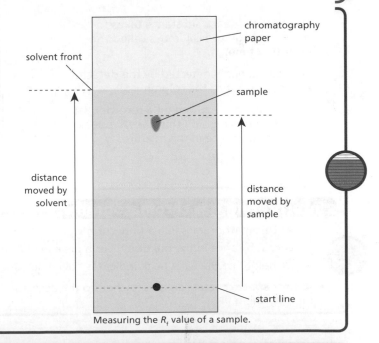

solvent front

chromatography paper

sample

distance moved by solvent

distance moved by sample

start line

Measuring the R_f value of a sample.

Improve your grade

Paper chromatography

Higher: Describe how to do a paper chromatogram to find out the R_f values of the dyes used to make an ink.

AO1 [4 marks]

Gas chromatography

How gas chromatography works

- Gas chromatography works in the same way as paper chromatography but uses a different stationary and mobile phase.

- Gas chromatography works for mixtures that turn into a gas easily when heated.

- The stationary phase is a solid instead of a piece of paper. The solid is packed into a long column. The column is coiled up so that it takes up less space.

- The substances are moved through the column by a gas. This is called the 'carrier gas' and is the mobile phase.

- The gas used must be inert so that it does not react with the substances being tested. Nitrogen or argon are often used.

- The column is surrounded by an oven. This heats up and controls the temperature of the column.

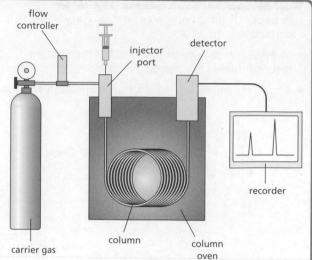

The arrangement of parts of a gas chromatography machine.

- The sample that is being tested contains a mixture of substances.
  - Some substances are attracted to the solid in the tube more than others.
  - Therefore each substance is carried along by the carrier gas at different speeds.
  - The substances spread out in the column, the ones that travel fastest come out of the end of the column first.

> ### EXAM TIP
>
> Make sure you can compare paper and TLC chromatography with gas chromatography. Gas chromatography uses a solid in a column instead of paper or a plate and uses a carrier gas instead of a solvent. Samples are liquid mixtures that turn into gases easily.

- The gas chromatogram gives a print out to show the time taken for each substance to travel through the column. This can be used to identify each substance.

Retention time

- Substances spread out in the column because some take longer than others to travel through to the end of the column.

- The time that each substance takes to travel through the column is called the **retention time**.

- Each substance is detected by the detector as it comes out of the column.

- The results are recorded on a recorder. The recorder gives a graph that shows a peak for each substance in the mixture.

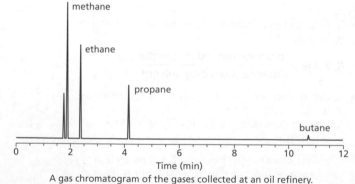

A gas chromatogram of the gases collected at an oil refinery.

Interpreting gas chromatograms

- A gas chromatogram is like a fingerprint for a mixture.
  - Every substance in the mixture gives a peak in a different place.
  - The height of the each peak indicates how much of each substance is in the mixture.

- A **mass spectrometer** can be attached to the detector. This uses the masses of the substances to identify them.

Improve your grade

Thin-layer and gas chromatography

Higher: Thin-layer chromatography and gas chromatography can both be used to separate and identify the substances in mixtures. Describe the differences between these two techniques.   *AO1* [4 marks]

Quantitative analysis

Using standard solutions in quantitative analysis

- A **standard solution** is a solution of known concentration.
- A standard solution contains a known amount of solute dissolved in a fixed volume of solution.

- Quantitative analysis involves taking measurements of quantities.
- Using standard solutions is important to reduce errors. Other procedures are also used to reduce errors.
- Measurements are repeated to try to check that there are no errors in the measurements.
- **Outliers** are measurements that do not fit the pattern of the other results. These can be left out of calculations if you are sure they are wrong.
- The **mean** of the results gives the **best estimate** of the **true value** of the quantity that is being measured.
- The **range** of results gives an idea of the **uncertainty** in the **accuracy** of the results.

Making standard solutions

- Standard solutions must be prepared carefully to make sure that their concentration is accurate.
  - Measure the mass of solute in a clean beaker.
  - Add distilled water and stir until it is fully dissolved.
  - Pour the solution into a **volumetric flask** using a funnel.
  - Rinse the beaker (and the glass rod if you have used one) and funnel with more distilled water and add the water to the flask. This makes sure that no solute 'escapes'.
  - Add more distilled water to the flask until it reaches the mark.
  - Invert the flask.

When a volumetric flask is filled up to the line it contains an accurately measured volume of liquid.

- The mass of the solute and the volume of the final solution are used to calculate the concentration of the standard solution.
- Concentration is measured in g/dm³ ('grams per litre' or 'grams per decimetre cubed').

EXAM TIP

Exam questions often ask how to make up a standard solution accurately. It is important to include detail, such as rinsing the beaker, adding the water dropwise near the end or reading from the bottom of the water meniscus.

Values in quantitative analysis

- To calculate concentration, the volume of the solution must be in dm³. If you have measurements in cm³ you need to start by converting the volume into dm³. You can do this by dividing by 1000 (because 1000 cm³ = 1 dm³).

Remember!

A result that is very different from the other results may not necessarily be an outlier. Many scientific discoveries have come from unexpected results!

- The concentration of a solution can be calculated using this formula.

$$\text{concentration of solution (g/dm}^3) = \frac{\text{mass of solute (g)}}{\text{volume of solution (dm}^3)}$$

- The amount of solute you need to use to make a particular concentration can be calculated by rearranging the formula like this.

$$\text{mass of solute (g)} = \text{concentration (g/dm}^3) \times \text{volume of solution (dm}^3)$$

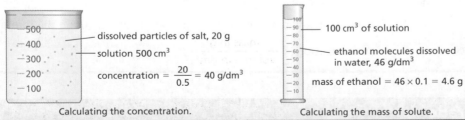

dissolved particles of salt, 20 g

solution 500 cm³

$$\text{concentration} = \frac{20}{0.5} = 40 \text{ g/dm}^3$$

Calculating the concentration.

100 cm³ of solution

ethanol molecules dissolved in water, 46 g/dm³

mass of ethanol = 46 × 0.1 = 4.6 g

Calculating the mass of solute.

Improve your grade

Calculating concentration

Higher: Harry made a standard solution of sodium chloride. He dissolved 10 g of sodium chloride in 250 cm³ of water.

What is the concentration of the solution?

AO2 [3 marks]

Acid-base titration

How to do titrations

- An acid-base titration measures exactly how much acid is needed to react with a base. The method makes sure that all measurements are as exact as possible.

- Measure out a known amount of base. A solid may need to be dissolved in water to make a standard solution. Use a pipette to measure the solution and put it into a conical flask.

- Add a few drops of indicator to the flask. A white tile under the flask helps to see the colour more clearly.

- Fill a burette with the acid. Run the acid through the burette to make sure the jet at the bottom is full. Take a start reading on the burette.

- Add the acid from the burette to the conical flask and swirl it to mix the solutions.

- Near the end point, add the acid drop by drop.

- When the indicator changes colour permanently, record the final reading on the burette.

- Treat the first reading as a 'rough' and do at least three more repeats to make sure that your readings are close together.

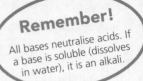

Remember!

All bases neutralise acids. If a base is soluble (dissolves in water), it is an alkali.

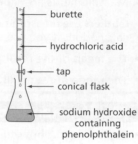

- burette
- hydrochloric acid
- tap
- conical flask
- sodium hydroxide containing phenolphthalein

Titration apparatus

- Titration results vary. You should aim for three accurate readings that are within 0.1 cm³ of each other. A small variation shows that the uncertainty in the results is small.

- There are many reasons why titration results vary. Most of these are caused by human error. For example:
  - Drips lost from the pipette or not filling the pipette accurately.
  - Solutions splashed or spilled.
  - Missing the colour change or adding the acid too quickly so that you miss the end point.
  - Misreading the burette.

Using titration results

- You can work out concentrations from titration results. You need to use the titration results and the relative formula mass (RFM) of each substance.

- For example:

  To find out the concentration of sodium hydroxide (NaOH) by using a titration with hydrochloric acid (HCl), the formula is:

$$\text{concentration of NaOH} = \frac{\text{concentration of HCl} \times \text{volume of HCl} \times \text{RFM NaOH}}{\text{RFM HCl} \times \text{volume of NaOH}}$$

EXAM TIP

In exam questions on foundation tier papers you will always be given the formula you need to do titration calculations.

- Titration calculations can also be worked out directly from the equation.

- Step 1: Work out the RFM of the substance in the flask (the base).

- Step 2: Work out the mass of the acid used in the titration. (You need to divide by 1000 to convert cm³ into dm³. Remember all concentration calculations are in dm³.)

$$\text{mass of acid used} = \frac{\text{concentration of acid} \times \text{volume of acid used in cm}^3}{1000}$$

- Step 3: From the equation, work out the mass of base that would exactly react with the mass of acid you have worked out in step 2. Remember to take into account the numbers in the equation that show how many RFM units react together.

- If the question asks for a concentration at the end, you may need to do a final calculation using the formula:

$$\text{concentration} = \frac{\text{mass in g}}{\text{volume in dm}^3}$$

Improve your grade

Titration calculation

Higher: May did a titration to find out the mass of calcium carbonate in a crushed indigestion tablet.

She used hydrochloric acid of concentration 15 g/dm³. She needed 25 cm³ of the acid to neutralise the calcium carbonate in the tablet.

$CaCO_3 + 2HCl \rightarrow CaCl_2 + H_2O + CO_2$

What is the mass of calcium carbonate in the tablet?

AO2 [4 marks]

C7 Summary

Bulk chemicals are made on a large scale, fine chemicals are made on a small scale.

The chemical industry

Chemists do research and development of new processes and products.

There are many laws about the chemical industry to protect people and the environment.

Chemical processes have a number of different stages.

Green chemistry

A number of different factors are involved in judging the sustainability of a process.

The atom economy of a process is a measure of the percentage of the mass of the atoms in the feedstock that end up in a product.

Exothermic reactions give out heat, endothermic reactions take in heat.

Energy changes

Energy changes may be shown using energy level diagrams.

Energy is taken in when bonds break and given out when bonds form.

Activation energy is used to break bonds to start a reaction.

Catalysts and enzymes increase the rate of reactions by providing an alternative reaction pathway with a lower activation energy.

Catalysts and enzymes

Enzymes are biological catalysts. They only work for specific reactions and work best in narrow ranges of temperature and pH.

Using enzymes can increase the sustainability of industrial processes because they reduce the energy needed but they only work under limited reaction conditions.

Bond energy values are different for different types of bond.

Energy calculations

The energy given out during a reaction can be calculated by working out the total energy taken in when all the bonds break in the reactants and the total energy given out when all the bonds form in the products.

Relative atomic mass is measured relative to a carbon atom with a mass of 12.

Reacting masses

Relative atomic masses are shown on the Periodic Table and can be used to work out relative formula masses.

Relative formula masses and an equation can be used to work out the masses of reactants and products involved in a reaction.

C7 Summary

Alkanes are hydrocarbons. Hydrocarbons contain carbon and hydrogen only.

When hydrocarbons burn they make carbon dioxide and water.

Alkanes

Alkanes burn but are generally unreactive because they have all single bonds that are difficult to break.

Saturated hydrocarbons contain single bonds between two carbon atoms. Unsaturated hydrocarbons contain double bonds between two carbon atoms.

Methanol and ethanol are both used as solvents and fuels. Ethanol is used to make alcoholic drinks.

Alcohols burn in air to make carbon dioxide and water in the same way as alkanes due to their hydrocarbon chain.

Alcohols

Alcohols contain the –OH functional group and therefore have different properties to alkanes.

Alcohols react more slowly than water with sodium.

During fermentation yeast turns sugar into dilute ethanol and carbon dioxide.

Fermentation and distillation

Yeast is a living microorganism which produces enzymes. Fermentation only works in a narrow temperature and pH range.

Concentrated ethanol must be made by distillation because a high concentration of ethanol is toxic to yeast.

Ethanol can be made from ethane from crude oil in a two-stage process.

Alternatives to fermentation

Ethanol can be made from biomass by using genetically modified *E.coli* bacteria.

Both processes have both advantages and disadvantages that affect their sustainability.

Methanoic acid and ethanoic acid are typical carboxylic acids with sharp tastes and strong smells.

Carboxylic acid

Carboxylic acids contain the functional group –COOH and react like other acids with metals, alkalis and carbonates.

When carboxylic acids react, the –COOH group loses H^+.

Strong acids are corrosive at all concentrations. Weak acids are not as hazardous because they are only corrosive at high concentrations.

Weak acids

Weak acids have a higher pH than strong acids and react more slowly with metals and carbonates.

When weak acids dissolve in water, most molecules in weak acids do not ionise so the concentration of hydrogen ions is lower.

C7 Summary

Esters

Esters have fruity smells and flavours. They occur naturally and are also made synthetically.

Esters are used as perfumes, flavourings, solvents and plasticisers.

Esters are made by heating together a carboxylic acid and an alcohol with a catalyst.

Making esters involves several stages; heating under reflux, distillation, purification using a tap funnel and drying.

Each stage uses different apparatus and has a different purpose.

Fats and oils

Fats come from animals and are solids at room temperature. Oils come from plants and are liquids at room temperature.

Fats and oils are esters made from glycerol and three fatty acid molecules.

Saturated fats and oils contain all single bonds between carbon atoms. Unsaturated oils contain some double bonds.

Bromine water goes orange/brown to colourless when it reacts with a molecule that contains a double carbon-carbon bond.

Reversible reactions

Some reactions are reversible and reach equilibrium in a closed container.

At equilibrium the amounts of the reactants and products stay the same because the rate of the forward and backward reactions are the same.

The Haber process

The Haber process fixes nitrogen by making ammonia from raw materials.

Changing the conditions of the Haber process affects the yield and the rate of the reaction.

The conditions chosen for the Haber process in industry are a compromise in terms of rate, yield and the cost of running the process.

Alternatives to Haber

Fertilisers cause environmental problems due to their large scale use.

New approaches to making fertilisers use more renewable resources and involve bacteria and enzymes.

Sustainable processes need to meet our demands for fertilisers without causing environmental harm.

C7 Summary

Qualitative analysis identifies the substances in a sample. Quantitative analysis gives numerical data about a sample.

Analysis

Samples collected for analysis must be chosen to represent the bulk material and need to be collected without contamination.

The conditions for storing samples for analysis need to be carefully controlled to limit the deterioration of the sample.

During chromatography the mobile phase moves and carries substances along. The stationary phase does not move.

Principles of chromatography

Substances can be distributed between solvents because they have different solubilities in different solvents.

The distance a substance travels is related to its distribution between the mobile phase and stationary phase.

During chromatography, a mixture is separated into different spots when a solvent travels up a paper or plate.

Locating agents are used to make colourless spots visible on a chromatogram.

Paper and thin layer chromatography

The R_f value of each spot can be calculated by dividing the distance the spot moves by the distance the solvent moves.

Substances can be identified by comparing the distance each spot moves to a reference sample.

Gas chromatography uses a carrier gas as a mobile phase and a solid in a column as a stationary phase to separate substances in a mixture of gases.

Gas chromatography

Different substances have different retention times. They are detected and recorded as they leave the column.

Print outs from the recorder can be used to identify substances and to indicate the amount of each substance in a mixture.

Standard solutions are made up accurately using a known mass of solute and a volumetric flask.

Quantitative analysis

When processing data it is important to consider outliers, estimates, true values, range, uncertainty and accuracy.

The concentration of a solution can be calculated by dividing the mass of solute in g by the volume of the solution in dm^3.

Titrations use accurate measuring equipment and a detailed procedure to give exact readings for the amounts of acid and alkali that react together.

Acid-base titration

Human and equipment error can cause small variations in titration results.

Titration results can be processed using a formula or an equation to calculate the masses or concentrations of unknown solutions.

C1 Improve your grade   Air quality

Page 3 The Earth's atmosphere

Foundation: The early atmosphere was mainly composed of water and carbon dioxide. Suggest how these were gradually removed.   *AO1* [4 marks]

Water condensed to form oceans. Carbon dioxide was removed by plant photosynthesis and by forming fossil fuels.

Answer grade: D/C. Both sentences are correct, but the answer lacks detail. For full marks, you need to explain that water condensed because the Earth cooled, and that fossil fuels form when plants and animals die and are buried under certain conditions. Carbon dioxide is also removed by dissolving in oceans and forming sedimentary rocks.

Page 4 How has human activity changed air quality?

Foundation: How do humans affect air quality?
          AO1 [4 marks]

Fuels such as coal give out waste gases. Cars give out gases that make asthma worse. Carbon dioxide is linked to acid rain. People cut down trees and burn them, adding carbon dioxide to the air.

Answer grade: E/D. This answer gains 1 mark. The first sentence gains a mark for linking pollutant gases to burning fuels, and the final sentence gains a mark for linking burning trees to carbon dioxide release. Sentence 2 is correct – cars emit both sulfur dioxide and nitrogen dioxide, which are linked to asthma – but this is not clearly stated. Sentence 3 has incorrect science. For full marks, you would need to correct sentences 2 and 3 and state that nitrogen oxides from cars are also linked to acid rain.

Page 5 What happens when fuels burn?

Foundation: Methane (CH_4) reacts with oxygen to form carbon dioxide and water.

Finish the diagram to show this reaction. Use ● to represent a carbon atom, ● to represent a hydrogen atom and ○ to represent an oxygen atom.

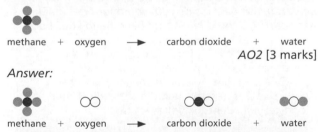

methane   +   oxygen   ⟶   carbon dioxide   +   water
          AO2 [3 marks]

Answer:

methane   +   oxygen   ⟶   carbon dioxide   +   water

Answer grade: C/D. The formula representations for oxygen, carbon dioxide and water are all correctly drawn, which gains 2 marks. However, the answer is incomplete. For full marks, the equation needs to be balanced. Do this by counting each atom and making sure that all of them are the same on both sides. To balance the equation, two molecules of water and two molecules of oxygen need to be drawn.

Page 6 How pollutants are formed

Foundation: Describe the damage acid rain causes, and explain why the UK is being blamed for acid rain damage in Europe.   *AO1* [4 marks]

Acid rain is made when sulfur dioxide dissolves in water vapour in clouds. The rain clouds are blown by winds. Acid rain kills trees and wildlife in lakes.

Answer grade: C/B. The first sentence uses correct scientific terms but does not explain the link between acid rain and pollution from power stations in the UK. The next two sentences state a fact but do not explain the causes of these facts. For full marks you need to explain why acid rain is a problem. For example, you could say that acid rain kills trees on land and wildlife in lakes by gradually changing the pH to such an extent that animals and plants cannot tolerate it.

Page 7 Reducing air pollution from transport

Foundation: Suggest ways of reducing air pollution from cars.   *AO1* [4 marks]

You could walk to school instead of getting someone to drive you. You could also take a bus instead of using a car. Finally, switch off your engine if you've stopped for a few minutes.

Answer grade: D/E. The first sentence gives a sensible suggestion and gains 1 mark. Sentence 2 is the same idea about not driving, so is too similar to get a separate mark. Sentence 3 is a different idea worth 1 mark. For full marks you would need to make some additional suggestions, such as make public transport cheaper and use cleaner fuels or clean exhaust gases.

Page 9 Comparing materials and measuring properties

Foundation: Sub aqua divers should always leave a marker buoy at the dive site to warn other boats to stay away, reducing the risk of divers being hit when they surface.

Describe the main properties that the marker buoy would need. *AO1 [4 marks]*

It will need to float and should be brightly coloured.

> **Answer grade: E.** This answer gives two sensible, simple properties, but these two ideas alone will not gain all the marks available for this question. For full marks, you need to suggest extra properties, such as being waterproof, hard-wearing or staying in shape.

Page 10 Natural and synthetic materials

Higher: What are synthetic materials, and what advantages do they have over natural materials?
AO1 [4 marks]

Some natural materials are found in the Earth's crust formed from plants and from animals. Limestone is a natural material than can be used to make a synthetic material. Synthetic materials are man-made. Cement is an example. It can be made to order but goes off if you leave it a long time or if it gets damp.

> **Answer grade: E/D.** Sentence 3 gains 1 mark, but while the other statements are true, they do not answer the question. For full marks, you would need to explain that synthetic materials are cheaper, can be made to order, are manufactured by chemical reactions, and enable properties to be designed to suit a particular purpose.

Page 11 Investigating boiling points

Higher: Explain why during the distillation of crude oil, small hydrocarbon molecules rise to the top of the tower. *AO1 [4 marks]*

The temperature drops with height up the tower. Small gases have lower boiling points. They can cool a lot more before becoming a liquid.

> **Answer grade: C.** All three sentences are worthy of 1 mark each. For full marks, you would need to add that small-sized molecules have weak intermolecular forces.

Page 12 Improving polymers

Higher: A company is making rotor blades for a radio-controlled toy helicopter.

A polymer needs to be made stronger but more flexible. What could be done, and how will it change the properties? *AO2 [3 marks]*

To make the plastic harder, cross-links could be made. To make it flexible, use a plasticiser. Plasticisers are small molecules inserted to disrupt the chains.

> **Answer grade D/C.** Sentences 2 and 3 are both correct and worth 1 mark. In Sentence 1, cross-linking will make the polymer harder, but this would prevent the plasticiser being added and would not make it flexible. For full marks, you need to add that increasing chain length to improve strength and using a plasticiser for flexibility would work.

Page 13 Nanotechnology

Foundation: Explain what nanoparticles are, and suggest why some act as catalysts. *AO1 [4 marks]*

Nanoparticles are small groups of atoms joined together. Some can be used as catalysts as they have a large area. They provide more sites for reactants to meet and react.

> **Answer grade: E/D.** Sentences 1 and 3 are worth 1 mark each. Sentence 2 is incorrect – it is not a large area. For full marks you need to add large surface area, and explain that nanoparticles work as catalysts by providing sites for reactants to meet. You would also need to include the idea of size – up to a thousand atoms or 100 nm.

Page 14 Are nanoparticles safe?

Higher: Fresh Crop is a company selling mixed salads. They are considering adding silver nanoparticles to their food packaging to help prevent bacterial decay.

Explain why some people believe this may have risks.
AO2 [3 marks]

The silver nanoparticles might get onto the food and be eaten. If it gets inside you, you might get ill. The nanoparticles could kill 'good' bacteria in your body.

> **Answer grade: D/C.** This answer is worth 2 marks for the ideas of getting inside and causing a possible problem. For full marks, the idea of uncertainty is needed, linked to the ideas that long-term evidence is not available.

C3 Improve your grade Chemicals in our lives

Page 16 Limestone, coal and salt

Foundation: Coal, limestone and salt are major raw materials for industry. Choose one and state how it is formed. *AO1* [4 marks]

Salt is made when water evaporates from the sea. Coal is made when dead plants decompose. Limestone is made from sediments.

Answer grade: F/G. The first sentence is worth 1 mark. The question says choose *one* raw material, so normally only the first answer will be marked. For full marks, you would need to state how salt is formed, i.e. warm sea drying up, rock dust combining, and being buried by sediments.

Page 17 Extracting and using salt

Higher: Describe how salt is obtained by solution mining, and suggest why the chemical industry prefers this method of extraction. *AO1* [4 marks]

Solution mining uses water to dissolve underground salt deposits. It comes out of the ground in solution, and this is easier for industry to use.

Answer grade: C. Each sentence is worth 1 mark each, but the answer lacks detail. For full marks, you need to explain that water needs to be pumped in under pressure, and salt solution returns up another pipe. Industry prefers this method as the salt obtained is purer, so there is less cost for removing impurities.

Page 18 About alkalis

Foundation: Use the two tables below to write a word equation for making potassium nitrate.

Soluble hydroxides	sodium hydroxide – NaOH
	potassium hydroxide – KOH
	calcium hydroxide – $Ca(OH)_2$
Soluble carbonates	sodium carbonate – Na_2CO_3
	potassium carbonate – K_2CO_3

Acid	Salt
hydrochloric – HCl	chloride – Cl
sulfuric – H_2SO_4	sulfate – SO_4
nitric – HNO_3	nitrate – NO_3

AO2 [4 marks]

Potassium hydroxide + sulfuric acid → potassium nitrate

Answer grade: E/F. This answer gains 1 mark for choosing potassium hydroxide as the alkali needed. For full marks, an acid needs choosing which will give nitrate, so the only one is nitric acid. Adding the equation: acid + base → salt + water would be useful, to show full understanding, and water needs to be added as a product.

Page 19 The benefits and risks of adding chlorine to drinking water

Foundation: Worldwide, over 100 000 people die from a disease called cholera each year, through drinking dirty water. In Britain, cholera is rare.

Explain these statements, and suggest advice for people who drink water directly from rivers. *AO2* [4 marks]

In the UK we add chlorine to water to kill cholera. The best advice for people in Third World countries is do not go to the toilet close to water.

Answer grade: D/E. Both sentences are worth 1 mark each. To improve to a C grade, you would need to add the following points: cholera breeds in dirty water; some countries do not have an enclosed water supply or cannot afford to chlorinate; the best advice is to boil drinking water.

Page 20 Should we worry about PVC?

Higher: Cling-film is used to wrap food. It may be made from thin sheets of PVC, which contain plasticiser molecules called phthalates to increase flexibility. These molecules have passed safety tests. Despite this, some people are still worried about their safety.

Suggest reasons for some people's reluctance to accept the risk. *AO2* [4 marks]

Phthalates might be able to get out of the cling-film on to the food. When the food is eaten, they might cause harm to your body. The safety test may not have been carried out correctly. If they think there is a risk, it's best to avoid it.

Answer grade: B/A. Sentences 1, 2 and 4 are worth 1 mark each. To improve to an A* grade, you could explain that safety tests might not test if phthalates can escape. You could also explain what phthalates might do to your body over time, and that it would be difficult to calculate the actual risk.

Page 23 The history of the Periodic Table

Foundation: Explain why Mendeleev's arrangement of elements was an improvement on Döbereiner's triads and Newlands' octaves. *AO1 [4 marks]*

Mendeleev's arrangement was better because it used the properties of elements and put them into groups. All of the element properties fitted, but elements in triads and octaves did not all fit. Triads and octaves only worked for some elements.

Answer grade: D/C. A good feature of this answer is that it talks about Döbereiner and Newlands, as the question asks. However, the student only discusses one aspect of the table – the idea that all of the element properties fit the table. The most important reasons that Mendeleev's table was an improvement are because he left gaps and he predicted the properties of new elements. When they were discovered, the 'missing' elements fitted Mendeleev's predictions.

Page 24 Finding elements in the Periodic Table

Higher: An atom has the electronic arrangement 2.8.1.

Identify the element and explain why its electronic arrangement shows that it is likely to be a metal. *AO2 [3 marks]*

The element is sodium. It is a metal because sodium is a metal.

Answer grade: C. The answer scores only 1 mark, for identifying the metal. You can do this by working out that the total number of electrons in the atom is 11, which is the same as the proton number of sodium. However, the answer does not explain what the electron arrangement shows. For the other 2 marks you would need to say that atoms with one electron in the outer shell are likely to be metals, and that they will be in Group 1, which only contains metals.

Page 25 Reactions of Group 1 elements with chlorine

Higher: Write the word and symbol equations for the reaction of sodium with bromine. Compare the rate of reaction of sodium and potassium with bromine. *AO1 [3 marks]*

sodium + bromine $\longrightarrow$ sodium bromide

$$Na + Br \longrightarrow NaBr$$

Potassium reacts faster because it is further down the group.

Answer grade: C/B. The word equation is correct but the formula for bromine is wrong – it should be Br_2. If you are aiming at grades A or B you need to be able to write equations for the reactions with bromine and iodine as well as chlorine. They follow the same pattern: just swap 'Br' for 'Cl' or 'I' in the equations. The correct equation is $2Na + Br_2 \longrightarrow 2NaBr$. The last point is correct, the reactivity increases down the group, so potassium reacts faster.

Page 26 Patterns in Group 7

Higher: Liz adds chlorine water to potassium bromide solution. The table shows what she sees and her explanation.

Halogen	Compound	Observations	Explanation
Chlorine	Potassium bromide	Solution turns brown	Bromine is made because chlorine displaces bromine. Chlorine is more reactive than bromine.

Predict what you will see when chlorine water is added to potassium iodide solution. Explain your reasoning. *AO2 [4 marks]*

You would see the solution go brown because iodine is made and it looks brown.

Answer grade: C/B. This answer gets 2 marks. The observations are correct, and it is correct that iodine is made, but you need to 'model' your answer on the explanation in the table. Look at the number of marks – there are 4 in total. To gain the other 2 marks available you need to mention that chlorine displaces iodine and explain that this is because chlorine is more reactive than iodine.

Page 27 Explaining properties

Foundation: Explain why sodium chloride conducts electricity when it is molten or dissolved in water but not when solid. *AO1 [4 marks]*

Sodium chloride conducts because it is an ionic compound and the ions need to move to be able to conduct electricity.

Answer grade: D. There are 4 marks available and several parts to the question, so you need to give an 'in-depth' answer here.

First, you need to explain why sodium chloride conducts electricity. This answer gains 1 mark by saying that sodium chloride is an ionic compound. However, this is a 'why' question so a higher-level answer is needed. The answer goes on to correctly say that the ions must be able to move, and gets 1 mark for this.

Notice that the question also asks about 'when molten' and 'when dissolved in water' and 'not when solid'. The answer has not mentioned any of these, so is only worth 2 marks. A better answer would go further to say that ions can only move when the compound is molten or when dissolved in water, but that ions cannot move in the solid.

Page 29 Simple molecular substances

Higher: The table shows some data about oxygen.

Boiling point	State at room temperature	Density	Electrical conductivity
–218 °C	gas	very low	does not conduct

Use ideas about bonding and forces between molecules to explain the properties of oxygen.

AO2 [3 marks]

Oxygen has a low boiling point and it is a gas. It has a low density and does not conduct. This is because it has covalent bonds.

Answer grade: C/B. This answer only gets 1 mark. The answer copies the information in the table (this is a common mistake) but does not explain any of the properties in terms of bonding and forces between molecules, which the question asks you to do. To gain full marks you need to link the low boiling point, density and state to the fact that the forces between molecules are very low, and the low electrical conductivity to the fact that the molecules have no charge.

Page 30 Testing for ions

Foundation: Sam tests a salt. He finds out it contains copper ions.

Describe what Sam does and what he sees.

AO2 [3 marks]

He adds some sodium hydroxide and he looks for a precipitate.

Answer grade: D/C. This answer makes two of the main three points. To get the last mark, it is important to make sure you give the colour of the precipitate.

Page 31 Giant covalent structures

Higher: Diamond is the hardest naturally occurring material on Earth.

Use ideas about the structure of diamond to explain why. *AO1* [3 marks]

Diamond is hard because the bonds are very strong. Each carbon atom in diamond is bonded to four others in a giant, 3-D structure, which holds each atom tightly in place.

Answer grade: B/A. This is a good answer which gains 2 of the 3 available marks. Notice how the answer uses these important terms – 'bonds', 'strong', 'giant structure' and '3-D'. However, a key term is missing from this answer – you need to describe the bonds as 'covalent'. It is very important to try to include the correct level of language when answering questions on the Higher tier.

Page 32 Using electrolysis

Higher: An electric current is passed through molten potassium chloride.

Write ionic equations to help you to explain what happens at each electrode and name the products that form. *AO1* [5 marks]

$2KCl \longrightarrow 2K + Cl_2$

The potassium chloride breaks down and it makes potassium and chlorine.

Answer grade: C. This response does not fully answer the question. The question does not ask for an overall equation, it asks for an equation at each electrode. The full answer needs to show each ionic equation: $K^+ + e \longrightarrow K$ and $2Cl^- \longrightarrow Cl_2 + 2e$. Also, saying that the potassium chloride 'breaks down' is not enough. You need to explain that the potassium ions are positive and so go to the negative electrode to gain electrons, and the negative chloride ions move to the positive electrode and lose electrons.

Page 33 Metals in the environment

Foundation: Old car batteries are made from lead. Lead is toxic. Modern batteries use alternative, non-toxic metals.

Use ideas about cost and benefit to explain why manufacturers could not stop using lead to make batteries until alternative batteries were developed.

AO3 [3 marks]

People could not go without cars so they needed to invent a new battery. Lead batteries are toxic but only the people at the garage touch them.

Answer grade: D/C. This answer gets 2 of the available 3 marks. It gives one benefit (needed for cars) and also talks about one cost to a group of people who were affected by using lead batteries. A better answer would identify more costs, such as the problems with dumping old car batteries or the environmental damage caused by toxic metals, or identify more benefits of continuing to use the old batteries, such as it is very expensive to develop new technology.

Page 35 Reactions of acids

Higher: Write a word equation and a balanced symbol equation to show the reaction between sodium hydroxide and sulfuric acid. *AO1 [3 marks]*

sodium hydroxide + sulfuric acid ⟶ sodium sulfate + water

$NaOH$ + H_2SO_4 ⟶ $NaSO_4$ + H_2O

Answer grade: C/B. This answer gains only 1 of the 3 available marks. The word equation is correct (hydroxides react with acids to give a salt and water), but the symbol equation is incorrect for two reasons. Firstly, the formula of the salt is incorrect: sodium ions have a +1 charge (Na^+) and sulfate ions have a –2 charge (SO_4^{2-}), so two Na^+ ions are needed to balance the charge of each SO_4^{2-} ion, i.e. the formula of sodium sulfate is Na_2SO_4. Secondly, the equation must be balanced. Counting the numbers of atoms on each side of the equation, the equation balances if a '2' is put in front of the NaOH, giving $2NaOH + H_2SO_4$ ⟶ $Na_2SO_4 + H_2O$.

Page 36 Reacting amounts

Foundation: What is the relative formula mass of calcium hydroxide, Ca(OH)$_2$?

Use the Periodic Table to help you. *AO2 [2 marks]*

RAM Ca = 39

RAM O = 16

RAM H = 1

Relative formula mass = 39 + 16 + 1 = 56

Answer grade: D. This gets 1 of the 2 available marks. It is a really good idea to write down the masses as you find them on the Periodic Table – this gets 1 mark. If you had made a small error you would still get 'error carried forward' if you have set your work out clearly. However, the relative formula mass is wrong. In the formula Ca(OH)$_2$ the 2 at the end shows that there are two oxygen atoms and two hydrogen atoms in the formula. So the correct answer is 39 + (2 × 16) + (2 × 1) = 73.

Page 37 Energy changes

Foundation: Sam does some experiments. She does four different reactions in test tubes and takes a note of the temperature before mixing and 60 s after mixing. Sam's results are shown in the table.

Experiment	Temperature before mixing (°C)	Temperature 60 s after mixing (°C)
1	17	21
2	18	11
3	16	18

Which reaction is the most exothermic? Explain how you can tell. *AO3 [3 marks]*

Reaction 2 because it is the biggest temperature change.

Answer grade: F. This answer is incorrect. Exothermic reactions give out energy, and in Experiment 2 the reaction is endothermic (takes in heat energy). For full marks you would need to say that the most exothermic reaction is Experiment 1, because this gives the largest temperature increase.

Page 38 Percentage yield

Higher: Ben prepares some copper sulfate crystals. He talks to Liz about his method.

Ben: 'First I added solid copper carbonate to sulfuric acid until the fizzing stopped.

Then I filtered off some unreacted copper carbonate.

Next I heated the filtrate to evaporate some of the water and left the solution to cool for a few minutes until some crystals formed.

Then I filtered off the crystals and weighed them. I made 1.4 g of crystals. I worked out that my theoretical yield is 1.6 g.'

Liz: 'Your percentage yield will be really inaccurate because you have missed some steps out of your method.'

Calculate Ben's percentage yield. Explain why his percentage yield is likely to be inaccurate. *AO2 [5 marks]*

$$Percentage\ yield\ = \frac{actual\ yield}{theoretical\ yield} \times 100\%$$

$$= \frac{1.4}{1.6} \times 100\% = 87.5\%$$

It will be inaccurate because Ben only did it once. He should have done it lots of times and taken an average. He didn't purify his crystals either.

Answer grade: B. This answer gets 3 of the available 5 marks. It gains 2 marks for correctly working out percentage yield. The point that Ben should have purified his crystals gains 1 mark. However, while repeating an experiment is a good idea, this takes too long when preparing salts, so this point does not score a mark.

Liz points out that the method has some missing steps – a better answer would identify these. To gain an additional 2 marks you could mention any two of the following steps: waiting for all the crystals to form; the use of crystallisation to purify the crystals; drying the crystals in an oven or dessicator; weighing the dried crystals.

Page 39 Changing rates of reactions

Foundation: Jack investigates the rate of the reaction between a large lump of calcium carbonate and dilute hydrochloric acid. He measures the volume of gas given off every 30 s. The reaction takes place very slowly.

Suggest what changes Jack could make to his experiment to make the reaction faster. *AO1 [3 marks]*

You could try heating it up because the reaction goes faster when it is at a higher temperature so that would make the gas come off faster.

Answer grade: E/D. This answer gets 1 mark for the idea of raising the temperature. However, if you look at the question, it clearly asks for *changes* – not just one change. There are 3 marks available, so to gain full marks you need to say three things. Two other changes Jack could make are: he could use a more concentrated acid, and he could use smaller pieces of calcium carbonate instead of one big lump.

Page 41 Transporting chemicals

Foundation: By law, tankers containing petrol must be clearly labelled with information for the fire service. Give reasons why this is necessary. *AO3 [3 marks]*

It is really important because then the fire service will know what to do if there is a fire.

> **Answer grade: F.** Answer only gains 1 of 3 marks. Use the number of marks as a guide to how many points to make in your answer, and notice that the question asks for 'reasons' – not just one. Other reasons could include:
> - In case of a spillage after an accident.
> - To protect the safety of people.
> - To limit harm to the environment after a spill.
> - So that the fire service would know how to clean up the spill.
> - To give information about what safety clothing was needed.
> - So that the fire service would know if the chemical was flammable/toxic.

Page 42 Making zinc sulfate

Higher: The two equations show different ways of making zinc sulfate.

Equation 1: $Zn + H_2SO_4 \rightarrow ZnSO_4 + H_2$

Equation 2: $ZnCO_3 + H_2SO_4 \rightarrow ZnSO_4 + H_2O + CO_2$

Which equation has the higher atom economy? Explain your reasoning. *AO3 [3 marks]*

Equation 1 because it only has one other product.

> **Answer grade: C.** Answer only gains 1 of 3 marks. A good answer should mention percentage and mass. Other points should include:
> - In equation 1 more of the total mass of the atoms in the reacting chemicals are used in the products.
> - Percentage mass of the atoms used is greater.
> - Less mass is wasted in by-products and waste.

Page 43 Energy changes and bonds

Higher: The reaction between hydrogen and chlorine is exothermic. The diagram shows molecules involved in the reaction.

$H-H + Cl-Cl \rightarrow H-Cl$
$ H-Cl$

Use ideas about energy changes and bonds to explain why this reaction is exothermic. *AO2 [3 marks]*

It is exothermic because more energy is needed to make bonds than to form them.

> **Answer grade: B.** This answer only gains 1 mark. It is actually mixed up. The candidate has made a very common error by using the word 'needed'. Energy is 'needed' implies that energy is taken in. This is only true when bonds are broken. Energy is not 'needed' when bonds form – forming bonds actually gives energy out. Also, the answer does not talk about the actual reaction at all. An answer that gained all 3 marks would need to say: *energy is taken in when bonds in hydrogen and chlorine break. Energy is given out when bonds form in hydrogen chloride. More energy is given out than taken in so the reaction is exothermic overall.*

Page 44 Catalysts and enzymes

Foundation: What are the similarities and differences between the way catalysts and enzymes work? *AO2 [4 marks]*

They both speed up reactions. Enzymes are biological catalysts and ordinary catalysts are not.

> **Answer grade: F.** The answer does not focus on what the question asks. It should include similarities and differences. For example:
> - Both provide a different route for the reaction.
> - Both lower the activation energy.
> 'Different' points include:
> - Catalysts are not sensitive to conditions.
> - Enzymes denature under a change in conditions *or*
> - pH and temperature may denature enzymes *or*
> - Enzymes can only be used in industry in a limited range of conditions.

Page 45 Using bond energies

Higher: Fluorine reacts with hydrogen to form hydrogen fluoride.

$H_2 + F_2 \rightarrow 2HF$

The bond energies for this reaction are shown in the table.

Bond	Energy (kJ/mol)
H-H	436
F-F	159
H-F	568

Calculate the energy change of reaction. *AO2 [3 marks]*

Bonds formed: 1H-F

Total energy given out = 568 kJ/mol

Bonds broken: H-H and F-F

Total energy taken in = 436 + 159 = 595 kJ/mol

Energy change of reaction = 1136 − 595 = 541 kJ/mol

> **Answer grade: A.** This answer gains 2 marks. In this case, the mistake is that only one H-F bond has been included in the calculation (look at the equation, there are 2HF molecules so two H-F bonds are made). However, the energy given out is correct, so is the overall calculation method, so 2 marks.

Page 46 Calculating masses

Higher: Hydrogen can be used as a fuel.

$$2H_2 + O_2 \rightarrow 2H_2O$$

What mass of oxygen is needed to completely burn 1 g of hydrogen? *AO2 [3 marks]*

Relative formula mass of H_2 = 2

Relative formula mass of O_2 = 32

The equation shows 2 g of hydrogen need 32 g of oxygen to burn.

Therefore 1 g of hydrogen needs 16 g of oxygen to burn.

Answer grade: A. This answer gains 2 of the 3 marks. The working is very well set out so that the examiner can give 'error carried forward' marks after one mistake. The answer shows the correct relative formula masses of hydrogen and oxygen (2 g and 32 g) and the candidate was right not to bother working out the mass of water (this is not mentioned in the question). However the candidate has not taken into account that the equation shows two molecules of hydrogen react with one molecule of oxygen. So the correct masses should be 4 g of hydrogen (2×2 g) react with 32 g of oxygen, leading to an answer of 8 g.

Page 47 Formulae of alkanes

Higher: The table shows information about the formulae of some alkanes.

Alkane	Number of carbon atoms	Formula
Ethane	2	C_2H_6
Propane	3	C_3H_8
Butane	4	C_4H_{10}

The next alkane in the series is called pentane. What is the formula of a molecule of pentane? Explain how you worked out your answer. *AO2* [3 marks]

The formula of pentane is C_5H_{12}.

I worked this out because the next alkane must have five carbon atoms. The hydrogen atoms go up by 2 every time another carbon atom is added.

Answer grade: A. This answer gains all 3 marks. Notice that the candidate has answered both parts of the question and has thought about making three points to access 3 marks. A very good answer.

Page 48 Reacting sodium with propanol

Higher: Propanol is an alcohol.

A piece of sodium is dropped into propanol and into water. Predict how the reaction of sodium with propanol will differ from the reaction of sodium with water. *AO3* [2 marks]

Both reactions will fizz. Propanol will react slower than water with sodium.

Answer grade: B. This answer gains 1 mark. The question asks how the reactions will differ, so the point that the candidate has made about the similarities (both will fizz) is not needed. Only one point has been made, although two points are available for the reaction.

Page 49 Making wine

Higher: Joe has a hobby making wine. He finds that one batch of wine has stopped fermenting. Suggest some reasons why the wine may have stopped fermenting. *AO1* [4 marks]

The wine might have stopped fermenting because the alcohol has killed the yeast. Alcohol is toxic to yeast.

Answer grade: C. This would gain 2 marks. The points made in the answer are correct but are not enough to access all 4 marks. Other reasons may be:
• There might not be enough sugar.
• The wine might be too hot or cold.
• Air might have got into the mixture.
• There may be a problem with the pH going out of the optimum range.

Page 50 Making ethanol

Higher: Describe the advantages and disadvantages of making ethanol from crude oil. *AO1* [3 marks]

Crude oil is not a good raw material to use for the environment because it will run out. We also need crude oil for other uses because we use it for making petrol. A lot of energy is needed to make ethanol from crude oil.

Answer grade: B. This answer gains 2 marks. Although the answer is well argued, contains lots of points and is very clear, the question asks for both advantages and disadvantages. This answer only gives disadvantages.

Page 51 Reactions of carboxylic acids

Higher: Complete the table to show the products of some reactions of carboxylic acids. *AO2* [3 marks]

Reactants	Products
methanoic acid + magnesium	magnesium methanoate + hydrogen
ethanoic acid + potassium hydroxide	potassium ethanoate + water
propanoic acid + sodium carbonate	sodium propanoate + water

Answer grade: B. This answer gains 2 marks. The first two rows are correct. For propanoic acid reacting with sodium carbonate the candidate has made two errors. The name has a missing 'o'. The name of the salt should be sodium propanoate. This is a very common error. Also, when acids react with carbonates they make carbon dioxide as well as water. This has been missed out.

Page 52 Reactions of strong and weak acids

Foundation: Describe how you use magnesium metal to tell the difference between a strong and a weak acid. Include what observations you would make. *AO2* [3 marks]

You would add magnesium to the acids and see which one reacted faster.

Answer grade: F. This answer gains 1 mark. Although the candidate clearly knows that the rate of reaction is different (the mark is for a mention of 'faster'), there are important parts of the answer missed out. The answer does not make it clear that the stronger acid would be faster. The question also asks for observations. A better answer would mention timing which one dissolved all of the magnesium first or which one fizzed faster.

Page 53 Nail varnish remover

Foundation: Ethyl ethanoate is used to make some types of nail varnish remover. Ethyl ethanoate is an ester. Explain the advantages of using ethyl ethanoate as a nail varnish remover. *AO2* [3 marks]

Esters have a fruity smell and so people like to use nail varnish remover because it smells good. Also esters are good solvents and so the nail varnish will come off easily if you use an ester for a nail varnish remover.

> **Answer grade: C.** This answer gains 3 marks. The candidate has covered both of the important points; esters smell nice and are good solvents. The answer also relates these properties to the advantages of using them in the nail varnish remover (people like the remover to smell nice and to remove the nail varnish easily). A good answer.

Page 54 Making ethyl ethanoate

Higher: Ethyl ethanoate is made from ethanoic acid and ethanol. Sulfuric acid is added to the mixture before it is heated. Explain why this is necessary. *AO2* [3 marks]

It is needed because it is a catalyst.

> **Answer grade: D.** This answer gains 1 mark. Just saying 'it is a catalyst' is not an explanation. For the remaining 2 marks, the answer needs to mention that the reaction is very slow at room temperature and that using a catalyst makes it much faster.

Page 55 Esters and oils

Higher: Nail varnish remover contains ethyl ethanoate. Face cream contains oils to soften the skin.

Give one similarity and one difference between the structure of a molecule of ethyl ethanoate and a molecule of oil. *AO2* [3 marks]

They are both esters. Ethyl ethanoate is a good solvent. Oils are not used as solvents.

> **Answer grade: C.** 1 mark. The question asks for a similarity and difference in the structure. The difference given is a use. Differences include the idea that the hydrocarbon chains in ethyl ethanoate are much shorter than in an oil, or that the alcohol used to make ethyl ethanoate (ethanol) has only one –OH group. The alcohol used to make an oil (glycerol) has three.

Page 56 Making sulfur trioxide

Higher: The equation shows the reaction between sulfur dioxide and oxygen to form sulfur trioxide. This reaction is important in the manufacture of sulfuric acid.

$$2SO_2 + O_2 \rightleftharpoons 2SO_3$$

Explain why this reaction cannot produce 100% yield of sulfur trioxide. *AO2* [3 marks]

It can't make 100% yield because it is a reversible reaction. This means that the reaction can go backwards so that the sulfur trioxide makes sulfur and sulfur dioxide again.

> **Answer grade: A.** This answer gains 2 marks. This is a well presented answer. It is very clear, and the answer refers to the names of the reactants and products (a very good idea to do this whenever you can). The only point that the answer does not mention is that the reaction reaches equilibrium.

Page 57 Saving waste

Higher: In the Haber process only about 20% of the hydrogen and nitrogen react in the reaction vessel to make ammonia.

Explain what is done to make sure that this does not result in a waste of feedstocks. *AO2* [3 marks]

There is no waste because the gases are recycled.

> **Answer grade: C.** This answer gains 1 mark. Three marks need three separate points to gain each mark. Possible points that could have been made include explaining how the feedstocks are not wasted:
> - The feedstocks for the process are hydrogen and nitrogen.
> - Hydrogen and nitrogen are recycled (not just 'gases' – this would include ammonia).
> - Ammonia is separated out.
> - The hydrogen and nitrogen are put back into the reaction vessel.
> - The cycle continues so that all of the hydrogen and nitrogen eventually react.

Page 58 Using fertilisers

Foundation: Give one advantage and one disadvantage of using synthetic fertilisers rather than organic fertilisers to grow food crops. *AO1* [2 marks]

One advantage of using synthetic fertiliser is that it makes crops grow quickly. A disadvantage is that it costs the farmer money.

> **Answer grade: U.** This answer gains no marks. The question asks for a comparison between synthetic and organic fertiliser. The two points that the candidate has made are actually true for both types of fertiliser. A better answer would be to give an advantage of synthetic fertiliser that is not true for organic fertiliser, e.g. *can be made more quickly in very large quantities* or *enough can be produced to meet the demands of farmers*. A disadvantage also needs to be a comparison, e.g. *synthetic fertilisers use more energy to make* or *synthetic fertilisers are more likely to wash into rivers and the sea*.

Page 59 Sampling tablets

Foundation: Carol works for a factory makes medicines in batches. The medicines are sold in bottles as tablets. The factory continues production for 24 hours every day. Carol's job is to check the contents of the tablets before they leave the factory.

How should Carol choose tablets to check to make sure that the checking is as thorough as possible? *AO2* [3 marks]

She should make sure she checks as many tablets as possible. It is really important that she makes sure that she doesn't contaminate the tablets and that she measures the amount of tablets she uses carefully.

> **Answer grade: E.** This answer gains 1 mark. The question asks about choosing tablets. The answer talks about contamination and measurement. There are many more points that could be made about how Carol should choose the tablets, e.g. from different batches, at different times of the day, from more than one bottle.

Page 60 Removing grease

Foundation: Dan has grease marks on his jeans. He washes them in a washing machine but the marks do not wash out. He uses a stain remover that contains a non-aqueous solvent to remove the marks. Explain why stain remover removes marks but washing does not. *AO2* [3 marks]

The marks were made of oil and water does not mix with oil so he couldn't wash the stains out just using water. He needed a stain remover.

Answer grade: D. This answer gains 2 marks. To gain the final mark, the answer should mention that different substances, such as oil, dissolve differently in different solvents, or say that oil is more soluble in non-aqueous solvents.

Page 61 Paper chromatography

Higher: Describe how to do a paper chromatogram to find out the R_f values of the dyes used to make an ink.

AO1 [4 marks]

Put a spot of ink on a chromatography paper and put the bottom of the paper in a solvent. When the solvent reaches nearly to the top of the paper, take the paper out. Measure the distance travelled by each spot and divide it by the distance travelled by the solvent to get the R_f value.

Answer grade: A. This answer gains all 4 marks. The candidate has answered all parts of the question; both how to do the chromatography and also what to do to calculate the R_f value at the end. More detail could have been put in, e.g. making sure the solvent is below the level of the spot, doing repeats and taking a mean, but these are not necessary in this case to gain all of the marks.

Page 62 Thin-layer and gas chromatography

Higher: Thin-layer chromatography and gas chromatography can both be used to separate and identify the substances in mixtures. Describe the differences between these two techniques.

AO1 [4 marks]

Thin layer chromatography uses a plate coated with a solid and a solvent. Gas chromatography uses a carrier gas to separate out the substances. You need an oven to heat up the samples in gas chromatography. It is easier to look at the results of gas chromatography because you get a print out from a recorder.

Answer grade: B. This answer gains 2 marks. If you are asked to give differences, it is wise to structure your answer to say 'one thing is like this and the other is like that' so that the comparison is clear. This answer gives information about gas chromatography but does not make the comparison to thin-layer chromatography very clear. The first two sentences show a comparison: use of a solvent rather than a gas. This gains 1 mark. A further mark can be gained for the points about gas chromatography (the oven, the use of a recorder) but clear comparisons need to be made to gain more than half marks. A better answer would mention that the solid on a plate for TLC is replaced by a solid in a column for gas chromatography, and that the results of TLC must be processed by comparison to reference samples or R_f values rather than from a print out. Other comparisons could include that TLC separates substances in solution rather than as a mixture of gases, or that mixtures do not vaporise easily.

Page 63 Calculating concentration

Higher: Harry made a standard solution of sodium chloride. He dissolved 10 g of sodium chloride in 250 cm³ of water.

What is the concentration of the solution?

AO2 [3 marks]

Concentration = 10 ÷ 250 = 0.04

Answer grade: C. This answer gains 1 mark. There are two errors here. Firstly, the candidate has forgotten to convert the volume given in the question from cm³ to dm³. The volume that should be used in the calculation is: 250 ÷ 1000 = 0.25 dm³. Secondly, if units are not given in the answer (sometimes a space with units already written in is given to you) then it is important that you include the units.

A fully correct answer is 10 ÷ 0.25 = 40 g/dm³

Page 64 Titration calculation

Higher: May did a titration to find out the mass of calcium carbonate in a crushed indigestion tablet.

She used hydrochloric acid of concentration 15 g/dm³. She needed 25 cm³ of the acid to neutralise the calcium carbonate in the tablet.

$$CaCO_3 + 2HCl \longrightarrow CaCl_2 + H_2O + CO_2$$

What is the mass of calcium carbonate in the tablet?

AO2 [4 marks]

RFM calcium carbonate, $CaCO_3$ = 40 + 12 + (3 × 16) = 100

mass of acid used $15 \times \dfrac{25}{1000} = 0.375$

In the equation.	$CaCO_3$	reacts with HCl
RFM	100 g $CaCO_3$	reacts with 36.5 g HCl
therefore	$\dfrac{100}{36.5} \times 0.375$ $CaCO_3$	reacts with 0.375 g HCl
	1.0 g $CaCO_3$	reacts with 0.375 g HCl

Tablet contains 1.0 g $CaCO_3$

Answer grade: A. This answer gains 3 marks. There is one error. Firstly, the answer is very well set out. This is important because if the final answer is not fully correct, the examiner can look back and give some marks for some correct working. If the answer is not well laid out it may be difficult for the examiner to see what you have done and you may lose marks. The answer starts by correctly working out the RFM of $CaCO_3$, for 1 mark.

Next, the mass of hydrochloric acid used in the titration is worked out, for 1 mark. Notice that the candidate has remembered to divide by 1000 to convert cm³ into dm³. The last part has a mistake. If you look at the equation, you will see that each RFM of $CaCO_3$ reacts with 2 HCl. In the working in the answer, the candidate has used a 1:1 ratio of 100 g $CaCO_3$ to 36.5 g HCl. As there are 2HCl in the equation, the ratio should be 100 g $CaCO_3$ to (36.5 × 2) = 73 g HCl. However, the rest of the answer is correct, so error carried forward can be given on the working to give 3 marks. The correct answer should be 0.5 g.

Ideas About Science

Understanding the scientific process

As part of your Science assessment, you will need to show that you have an understanding of the scientific process – Ideas about Science.

Science aims to develop explanations for what we observe in the world around us. These explanations must be based on scientific evidence, rather than just opinion. Scientists therefore carry out experiments to test their ideas and to develop theories. The way in which scientific data is collected and analysed is crucial to the scientific process. Scientists are sceptical about claims that cannot be reproduced by others.

You should be aware that there are some questions that science cannot currently answer and some that science cannot address.

Collecting and evaluating data

You should be able to devise a plan that will answer a scientific question or solve a scientific problem. In doing so, you will need to collect and use data from both primary and secondary sources. Primary data is data you collect from your experiments and surveys, or by interviewing people.

While collecting primary data, you will need to show that you can identify risks and work safely. It is important that you work accurately and that when you repeat an experiment, you get similar results.

Secondary data is found by research, often using ICT (the Internet and computer simulations), but do not forget that books, journals, magazines and newspapers can also be excellent sources. You will need to judge the reliability of the source of information and also the quality of any data that may be presented.

Presenting and processing information

You should be able to present your information in an appropriate, scientific manner, using clear English and the correct scientific terminology and conventions. You will often process data by carrying out calculations, drawing a graph or using statistics. This will help to show relationships in the data you have collected.

You should be able to develop an argument and come to a conclusion based on analysis of the data you collect, along with your scientific knowledge and understanding. Bear in mind that it may be important to use both quantitative and qualitative arguments.

You must also evaluate the data you collect and how its quality may limit the conclusions you can draw. Remember that a correlation between a factor that's tested or investigated and an outcome does not necessarily mean that the factor caused the outcome.

Changing ideas and explanations

Many of today's scientific and technological developments have benefits, risks and unintended consequences.

The decisions that scientists make will often raise a combination of ethical, environmental, social and economic questions. Scientific ideas and explanations may change as time passes, and the standards and values of society may also change. It is the job of scientists to discuss and evaluate these changing ideas, and to make or suggest changes that benefit people.

Glossary

A

accuracy how near a reading is to the true value 20, 63

acid a chemical compound which when dissolved in water gives a pH reading of under 7 and turns litmus red 4, 5, 6, 18, 21, 30, 34–40

acid rain rainwater which is made more acidic by pollutant gases 4–6, 8, 33

activation energy the minimum amount of energy needed by the reactants for a reaction to take place 43, 44, 45

alcohol a family of organic compounds containing an OH group, for example, methanol (CH_3OH); the common name for ethanol (C_2H_5OH) 48

alkali a chemical compound which when dissolved in water gives a pH reading of over 7 and turns litmus blue 18, 21, 25–26, 30, 35–37, 40

alkali metal very reactive metal in Group 1 of the Periodic Table, for example sodium 25–26

alkanes a family of hydrocarbons (C_nH_{2n+2}) found in crude oil 10, 47

anode positive electrode 19, 32

aqueous dissolved in water 60

atom the basic 'building block' of an element which cannot be chemically broken down 5–6, 8–9, 11–14, 23–36

attractive a force that pulls two objects together 11, 12, 29, 33

B

background radiation low-level radiation that is found all around us 11, 12, 29, 33

bacteria single-celled microorganisms, some of which may invade the body and cause disease 3,14–15, 17, 28

balanced symbol equation a symbolic representation showing the kind and amount of the starting materials and products of a chemical reaction 18, 25, 35

base solid alkali; any substance that neutralises an acid 18, 21

best estimate the closest you can get to the true value of a quantity from a set of data: it is usually the mean of a set of data 63

biodegradable a material that can be broken down by microorganisms 19

biofuel fuel such as wood, ethanol or biodiesel, obtained from living plants 7–8

biomass the amount of organic material present in an ecosystem, such as a pond; also, the amount of organic material in an organism (usually measured as dry mass). Also, plant material (often waste from other uses) that is used as a source of energy, for example, through burning in an electricity generator 50

bulk chemical a chemical that is manufactured and used in large amounts, in excess of 1 million tonnes a year, for example, sulfuric acid 41

by-product a product of a reaction that is not the main useful product. By-products may be useful or not 42

C

carbon an element that combines with others, such as hydrogen and oxygen, to form many compounds in living organisms 3–8, 10–15, 18, 20–21, 26–27, 29–32, 34–36

carbon dioxide gas whose molecules consist of one carbon and two oxygen atoms, CO_2; product of respiration and combustion; used in photosynthesis; a greenhouse gas 3–8, 10, 18, 29–31, 34–35

carbon monoxide poisonous gas whose molecules consist of one carbon and one oxygen atom, CO 4, 6, 8

carboxylic acids a family of organic compounds with the – COOH functional group 51

catalyst chemical that speeds up a chemical reaction but is not itself used up 7, 13, 39–40, 44

cathode negative electrode 19, 32

ceramics non-metallic solids made by heating and cooling a material, such as clay to make pottery 10, 12

chemical synthesis combining simple substances to make a new compound 35–40

combustion process in which substances react with oxygen releasing heat 5–6

composite material consisting of a mixture of other materials 14

compound substance composed of two or more elements which are chemically joined together, for example H_2O 5–6, 10, 18, 25–32, 34–36, 40

compressive strength a measure of resistance to squeezing or crushing forces 9

condense to turn from a gas into a liquid, as in steam (water vapour) which condenses to liquid water 8, 11

conductor a substance in which electric current can flow freely 10, 33

correlation a link between two factors that shows they are related, but one does not necessarily cause the other; a positive correlation shows that as one variable increases, the other also increases; a negative correlation shows that as one variable increases, the other decreases 4

corrosive a substance that can destroy or eat away other substances by a chemical reaction, e.g. it will burn skin 25–26

covalent bonds these join together the atoms inside a molecule 29, 31, 34

crust surface layer of Earth, made up of tectonic plates 10, 14, 16, 21, 31

crystal lattice crystals formed by ionic compounds, such as sodium chloride, which have a regular repeating pattern and shape 27, 30

crystallise when a liquid undergoes evaporation the product left behind cools and starts to form crystals 38

crystals the solid residue left after salts evaporate, they have regular shapes and flat sides 30, 34, 38, 40

current flow of electrons in an electric circuit 19, 26, 32, 34

D

data information, often in the form of numbers obtained from surveys or experiments 4–6, 9, 20, 23, 29, 33, 36

decompose in chemistry, separation of a chemical compound into simpler compounds 32

denature an enzyme is denatured if its shape changes so that it is no longer able to act as a catalyst 44

density the mass of a substance per unit volume 9–11, 14, 25, 29

diatomic molecules atoms that are joined together in pairs 26, 28

displacement the distance moved in a specific direction 26, 28

displacement reactions the difference in the reactivity of halogens. Where one halogen will take the place of another in its compounds 26, 28

dissolve to be soluble in water 3, 5–6, 8, 30–32, 34–35, 37–38

dissolving the act of a solid mixing into a liquid to form a solution 21, 38

distribute spread (of a solute) between two immiscible solvents 60

dry air air that has had all water vapour removed 3, 29, 34

E

effervescence the fizzing and bubbling effect that occurs e.g. when an acid reacts with a carbonate ion 30

elastic a material that returns to its original shape and size after a deforming force is removed 9

electric current a negative flow of electrical charge through a medium, carried by electrons in a conductor 19, 32, 34

Glossary

electrodes solid electrical conductors through which the current passes into and out of the liquid during electrolysis – and at which the electrolysis reactions take place 27

electrolysis decomposing an ionic compound by passing an electric current through it while molten or in solution 19, 21, 32, 34

electrolyte the liquid in which electrolysis takes place 32

electron tiny negatively charged particle within an atom that orbits the nucleus – responsible for current in electrical circuits 23–25, 27–29, 31–34

electron arrangement the configuration of electrons in shells, or energy levels, in an atom 24, 27–28

element substance made out of only one type of atom 5–6, 8, 10, 18, 20, 23–29, 31–32, 36

endothermic a reaction that reduces the temperature of the surroundings. The temperature falls in endothermic reactions 37, 40, 43

end-point the sudden change of colour of an indicator, e.g. in titration 36

energy level describes the arrangement of electrons in an atom in shells 37, 40

energy level diagram visual way of showing the change in energy level during a chemical reaction 37, 40, 43

environment an organism's surroundings 4, 14, 17, 19–21, 29–34, 39

enzyme a substance (usually protein) produced in cells that act as catalysts in reactions. Enzymes control many of the processes in cells but can also be used as catalysts outside cells 44

equilibrium the state of a reversible reaction when the amount of reactants and products remains constant 56

erosion the wearing away of rock or other surface matter such as soil 21

error uncertainty in scientific data 9, 36

ethanoic acid a carboxylic acid (CH_3COOH) found in vinegar; also known as acetic acid 51

evaporate turn from a liquid to a gas, such as when water evaporates to form water vapour 11, 16, 30, 38

exothermic a reaction that gives out heat to the surroundings. The temperature rises in exothermic reactions 37, 40, 43

F

fatty acids carboxylic acids with a long hydrocarbon chain; the chain may be saturated or unsaturated 55

feedstock the chemicals used in a manufacturing process 42

fermentation a process in which sugar is converted into ethanol (alcohol) by the action of enzymes, such as in yeast 49

fibre a long thin thread or filament 9–12, 13, 15

field in physics, a space in which a particular force acts 16

filtrate the insoluble products that remain trapped in a filter 38

filtration a method of separating one substance out from others. Filtering separates solids from liquids 38, 40

fine chemical A chemical that is manufactured and used in small amounts, up to a few tonnes a year, for example, drugs and flavourings 41

fixing (nitrogen) a process in which nitrogen gas in the air is converted into soluble compounds in the soil that can be used by plants 57

flue gas desulfurisation industrial process whereby sulfur is removed from waste gases 7–8

force the push or pull that acts between two objects 9, 11–12, 15, 29–31, 34

formula (for a chemical compound) group of chemical symbols and numbers, showing elements, and how many atoms of each, a compound is made up of 6, 10, 25–32, 34–37, 40

fossil fuel fuel such as coal, oil or natural gas, formed millions of years ago from dead plants and animals 6–8

fraction group of substances with similar boiling points, produced by fractional distillation 11, 15

fractional distillation process that separates the hydrocarbons in crude oil according to size of their molecules 11, 15

free electrons the outer electrons of atoms of materials that are good conductors which are loosely held and can break free easily so they can move freely 33

functional group the part of a molecule that is responsible for the reactions of the molecule, for example, the OH group in alcohols 48

G

gas state of matter in which atoms or molecules are spaced far apart and spread out to fill the available space 3–8

geologist scientist who studies rocks and the changes in the Earth 16, 21

giant covalent structures an element made with very strong covalent bonds between atoms in which a large number of carbon atoms are linked together in a regular pattern 31, 34

glycerol an organic compound with three carbon atoms, each with the alcohol functional group (–OH); glycerol combines with fatty acids to make esters that are fats and oils 54

gradient the degree of slope of a line 39

gram formula mass the number of grams of an element or compound represented by its RAM or RFM 32, 34

groups within the Periodic Table, the vertical columns are called groups 23

H

Haber process an industrial process for making ammonia 57

habitat the physical surroundings of an organism 17, 33

halides compounds formed when halogens react with alkali metals and other metals 26

hardness a measure of resistance to change in shape of a solid, for example by scratching or by impact 9, 15

hazard something that is likely to cause harm, e.g. a radioactive substance 17, 25, 35

high blood pressure blood pressure that is consistently abnormally high 17

hydrocarbon compound containing only carbon and hydrogen 5, 10–11, 15, 47

hydrosphere made up of the water, ice and snow on the Earth's surface and the water vapour in the atmosphere 30

I

inert an element that does not react with any other elements 24

igneous rock rock formed by the solidification of molten magma or lava 16

immiscible liquids that do not mix, but form separate layers, are immiscible 60

indicator in chemistry, a substance that shows the presence of an acid or an alkali by a change in colour; in biology, a measure of the quality of a natural environment, for example, the number of sensitive species present in an aquatic environment, or the level of pollutants in the air 18, 25, 35–36, 40

insoluble not soluble in water (forms a precipitate) 3, 5, 17–18, 21, 30

instantaneous speed the speed at a particular moment in

ion atom (or groups of atoms) with a positive or negative charge, caused by losing or gaining electrons 25, 27–28, 30–35, 37, 40

ionic bond chemical bond between two ions of opposite charges 30

ionic compounds salts made up of particles called ions which have a positive or negative electrical charge 27–28, 31, 33–35

ionic equation a chemical equation that describes changes that occur in aqueous solutions 30, 32, 34, 37, 40

Glossary

L

lattice a repeating pattern formed by the regular 3-D arrangement of ions 27, 30–31, 34

lava molten rock (magma) from beneath the Earth's surface when it erupts from a volcano 3, 16

Life Cycle Assessment an analysis of the environmental impact of a product, including the production of raw materials, its manufacture, packing, transport, use and disposal 20–21

line spectrum a set of different coloured lines produced when the light from a burning element is passed through a prism 23, 28

lithosphere the rocky outer section of Earth, consisting of the crust and upper part of the mantle 31

locating agent a chemical used to show up the position of colourless materials in paper or thin layer chromatography 61

M

magnetic field a space in which a magnetic material exerts a force 16

malleable able to be beaten into a thin sheet; a common property of metals 10, 33

mantle semi-liquid layer of the Earth beneath the crust 31

mass spectrometer an instrument that can be used to identify compounds by a comparison of the abundance and mass of fragments of the molecules 62

mean an average of a set of data 6, 9, 11, 15, 36, 63

melting point temperature at which a solid changes to a liquid 9, 12, 15, 24–25, 26–27, 29, 33–34

membrane (of a cell) the layer around a cell which helps to control substances entering and leaving the cell 19

membrane cell electrolysis cell that uses a semi-permeable membrane to separate the reactions at the two electrodes, as in the electrolysis of brine 19

metal a group of materials (elements or mixtures of elements) with broadly similar properties, such as being hard and shiny, able to conduct heat and electricity, and able to form thin sheets (malleable) and wires (ductile) 3, 5, 8, 10, 14–15, 18, 24–35, 40

metallic bond the force in metals that attracts atoms together 33

methane a gas with molecules composed of carbon and hydrogen; a greenhouse gas 3, 5, 10, 29

methanoic acid a carboxylic acid (HCOOH), found in ant and nettle stings; also known as formic acid 51

microorganism very small organism (living thing) which can only be viewed through a microscope 14, 19

minerals solid metallic or non-metallic substances found naturally in the Earth's crust 31, 34

mixture one or more elements or compounds mixed together but not chemically joined, so they can be separated out fairly easily 3, 6, 10–11

mobile phase the moving part of a chromatography system, consisting of a liquid or gas, which carries the components of the mixture at varying speeds 60

molecular ion a charged ion composed of two or more atoms joined together by covalent bonds 30

molecule two or more atoms held together by strong chemical bonds 3, 5–6, 8, 10–13, 15, 20–21, 26, 28–29, 31, 34, 37

monoculture when a single crop is grown 5

monomers small molecules that become chemically bonded to one another to form a polymer chain 11, 15

N

nanometre (nm) unit used to measure very small things (one-billionth of a metre, or 10^{-9}m) 13

nanoparticles very small particles on an atomic scale 13–15

nanotechnology technology making use of nanoparticles 14

nanotube a carbon molecule in the form of a cylinder 13–14

natural materials materials made from plant and animal products 10, 13, 15

neutral in chemistry the term neutral means 'between acid and alkali' 7–8, 18, 32, 35–37

neutralisation reaction between an acid and a base (H+ ions and OH– ions), to make a salt and water 18, 36–37, 40

neutron small particle that does not have a charge – found in the nucleus of an atom 23, 28

nitrogen oxides gaseous molecules containing nitrogen and oxygen atoms according to the formula NO_x, where X = 1, 2, etc.; these pollutants are formed due to the high temperatures created by the combustion of fossil fuels 14, 16, 18

non-aqueous a liquid consisting of a compound that is not water-soluble 60

nucleus the central core of an atom, which contains protons and neutrons and has a positive charge 23–24, 28–29

O

orbits electrons are arranged in orbits (or shells) around the nucleus of an atom 23

ores rocks that contain minerals, including metals, e.g. iron ore 31, 34

outlier a measurement that does not follow the trend of other measurements 6, 9, 15, 36, 63

oxidation chemical process that increases the amount of oxygen in a compound; the opposite of reduction 5, 18, 31, 34

oxidised a substance that has undergone oxidation 7, 11, 31

ozone gas found high in the atmosphere which absorbs ultraviolet rays from the Sun 19

P

particulates pollution in the form of particles in the air, such as soot 4, 7, 13

period a horizontal row in the periodic table 24, 28

Periodic Table a table of all the chemical elements based on their atomic number 23–28, 32, 40

pH a measure of the acidity or alkalinity of a substance 5, 18, 25, 35, 37, 40, 52

plastic a compound produced by polymerisation, capable of being moulded into various shapes or drawn into filaments and used as textile fibres 9–12, 14–15, 19–21

plasticiser small molecules which fit between polymer chains and allow them to slide over each other 12, 15, 20–21

pollutant harmful substance in the environment 3–8, 33

polymer large molecule made up of a chain of smaller molecules (monomers) 10–12, 15

polymerisation chemical process that combines monomers to form a polymer 11

precipitate insoluble solid formed in a solution during a chemical reaction 30, 34

products chemicals produced at the end of a chemical reaction 5–10, 18–21, 32, 36–38, 40

proton small positively charged particle found in the nucleus of an atom 23–24

proton number the number of protons in an atom 23–24, 28

pure a substance that has nothing else mixed with it 5, 29, 33, 35, 38

PVC a type of polymer (short for polyvinyl chloride) 12, 19–21

Q

qualitative data data that describes or depends on a property or characteristic, such as colour, not expressed using numbers 59

quantitative data numerical data 59

Glossary

R

radioactive a material that randomly emits ionising radiation from its atomic nuclei 16

radioactive decay the disintegration of a radioactive substance, the process by which an atomic nucleus loses energy 16

range in a series of data, the spread from the highest number to the lowest number 63

rate a measure of speed; the number of times something happens in a set amount of time 25, 39–40

rate of reaction the speed with which a chemical reaction takes place 25, 39–40

reactants chemicals that react together in a chemical reaction 5, 10, 32, 35–40

redox reaction the reaction for extracting metals from their ores, involving both oxidation and reduction 31

reduces when the atoms of a substance are oxidised resulting in the reduction of another substance 31

reduction process that reduces the amount of oxygen in a compound, or removes all the oxygen from it – the opposite of oxidation 5, 31, 34

reference materials known substances used for comparison in analysis, particularly in chromatography 61

reflux a process in which reactants can be kept at boiling point by condensing gases that are formed and returning them to the reaction vessel 54

relative atomic mass (RAM) the mass of an atom compared to the mass of an atom of carbon (which has a value of 12) 23, 32, 34, 36, 40, 46

relative formula mass (RFM) the sum of the RAMs of all the atoms or ions in a compound 32, 34, 36, 40, 46

reproducibility the ability of the results of an experiment to be reproduced by another experimenter 9

retention time the time taken for a chemical to pass through a gas chromatography column 62

reversible a reaction that can go both forwards and backwards 56

risk the likelihood of a hazard causing harm 14, 16–21, 29, 35

S

salt generically, the dietary additive sodium chloride; in chemistry, an ionic compound formed when an acid neutralises a base 3, 13, 16–18, 21, 23, 27, 30, 34–35, 37, 40

saturated a molecule that contains only single carbon-to-carbon covalent bonds 47

seafloor spreading an extension of the seafloor caused by tectonic plate movement and the extrusion of magma between two plates which solidifies to form rock 3

sediment particles of rock etc. in water that settle to the bottom 3, 8, 16

sedimentary rock rock formed when sediments are laid down and compacted together 3, 8, 16, 21

sedimentation the settling of particles in water to the bottom 21

shells electrons are arranged in shells (or orbits) around the nucleus of an atom 23–24

smog air pollution that is caused, for example by vehicle emissions and industrial fumes 6

soluble able to dissolve (usually in water) 18, 27, 30, 38

solvent a liquid in which solutes dissolve to form a solution 53

standard solution a solution with a known concentration of solute 63

state symbols symbols used in equations to show whether something is solid (s), liquid (l), gas (g) or in solution in water (aq) 30

stationary phase the part of a chromatography set up which does not move, consisting of a solid or liquid to which components of a mixture can stick to a greater or lesser extent 60

stiffness a measure of the resistance of a solid to bending forces 9

sulfur dioxide pollutant gas released from burning sulfur-containing fuels, which causes acid rain 4–7, 33

supercontinent very large land mass 16

sustainable a resource or process that will still be available to future generations 42

synthesis the building up of larger molecules through chemical reactions 42

synthetic material material manufactured from chemicals 10

T

tap funnel apparatus for separating liquids that do not mix together, consisting of a container with a tap at the bottom 54

tarnish the reaction that occurs when a metal reacts with oxygen in the air 25

tensile strength a measure of the resistance of a solid to a pulling or stretching force 9

theory a creative idea that may explain an observation and that can be tested by experimentation 3, 16

thermoplastic plastic with a shape that can be changed by heating 12

thermosetting plastic with a shape that becomes permanent after heating and cooling 12

titration common laboratory method used to determine the unknown concentration of a known reactant 36

toxic a poison or hazardous substance that can cause serious medical conditions or death 26, 81

trend the changes in a property across a period of the Periodic Table 24, 26

true value a theoretically accurate value that could be found if measurements could be made without errors 36, 63

U

uncertainty the interval within which the true value can be expected to lie, with a given level of confidence or probability 63

unsaturated a molecule that contains one or more carbon-to-carbon double covalent bonds 47

V

volcano landform from which molten rock erupts onto the surface 3

volumetric flask A glass vessel with a long neck that has a gradation marked on it; when the volumetric fl ask is filled to the mark it contains a known volume of liquid to considerable accuracy 63

W

weak acid an acid that reacts more slowly and has a higher pH than a strong acid with the same concentration; only partly split up into ions 52

Data sheet

Tests for negatively charged ions

Ion	Test	Observation
carbonate CO_3^{2-}	add dilute acid	effervesces, and carbon dioxide gas produced (the gas turns lime water milky)
chloride (in solution) Cl^-	acidify with dilute nitric acid, then add silver nitrate solution	white precipitate
bromide (in solution) Br^-	acidify with dilute nitric acid, then add silver nitrate solution	cream precipitate
iodide (in solution) I^-	acidify with dilute nitric acid, then add silver nitrate solution	yellow precipitate
sulfate (in solution) SO_4^{2-}	acidify, then add barium chloride solution or barium nitrate solution	white precipitate

Tests for positively charged ions

Ion	Test	Observation
calcium Ca^{2+}	add sodium hydroxide solution	white precipitate (insoluble in excess)
copper Cu^{2+}	add sodium hydroxide solution	light blue precipitate (insoluble in excess)
iron(II) Fe^{2+}	add sodium hydroxide solution	green precipitate (insoluble in excess)
iron(III) Fe^{3+}	add sodium hydroxide solution	red-brown precipitate (insoluble in excess)
zinc Zn^{2+}	add sodium hydroxide solution	white precipitate (soluble in excess, giving a colourless solution)

Formulae of some common molecules and compounds*

H_2	hydrogen gas	CH_4	methane	KCl	potassium chloride
O_2	oxygen gas	NH_3	ammonia	MgO	magnesium oxide
N_2	nitrogen gas	H_2SO_4	sulfuric acid	$Mg(OH)_2$	magnesium hydroxide
H_2O	water	HCl	hydrochloric acid	$MgCO_3$	magnesium carbonate
Cl_2	chlorine gas	HNO_3	nitric acid	$MgCl_2$	magnesium chloride
CO_2	carbon dioxide	NaCl	sodium chloride	$MgSO_4$	magnesium sulfate
CO	carbon monoxide	NaOH	sodium hydroxide	$CaCO_3$	calcium carbonate
NO	nitrogen monoxide	Na_2CO_3	sodium carbonate	$CaCl_2$	calcium chloride
NO_2	nitrogen dioxide	$NaNO_3$	sodium nitrate	$CaSO_4$	calcium sulfate
SO_2	sulfur dioxide	Na_2SO_4	sodium sulfate		

* These will not be provided in your exam. You need to learn them.

Key

1
H
hydrogen
1

relative atomic mass
atomic symbol
name
atomic (proton) number

1	2												3	4	5	6	7	0
																		4 He helium 2
7 Li lithium 3	9 Be beryllium 4												11 B boron 5	12 C carbon 6	14 N nitrogen 7	16 O oxygen 8	19 F fluorine 9	20 Ne neon 10
23 Na sodium 11	24 Mg magnesium 12												27 Al aluminium 13	28 Si silicon 14	31 P phosphorus 15	32 S sulfur 16	35.5 Cl chlorine 17	40 Ar argon 18
39 K potassium 19	40 Ca calcium 20	45 Sc scandium 21	48 Ti titanium 22	51 V vanadium 23	52 Cr chromium 24	55 Mn manganese 25	56 Fe iron 26	59 Co cobalt 27	59 Ni nickel 28	63.5 Cu copper 29	65 Zn zinc 30		70 Ga gallium 31	73 Ge germanium 32	75 As arsenic 33	79 Se selenium 34	80 Br bromine 35	84 Kr krypton 36
85 Rb rubidium 37	88 Sr strontium 38	89 Y yttrium 39	91 Zr zirconium 40	93 Nb niobium 41	96 Mo molybdenum 42	[98] Tc technetium 43	101 Ru ruthenium 44	103 Rh rhodium 45	106 Pd palladium 46	108 Ag silver 47	112 Cd cadmium 48		115 In indium 49	119 Sn tin 50	122 Sb antimony 51	128 Te tellurium 52	127 I iodine 53	131 Xe xenon 54
133 Cs caesium 55	137 Ba barium 56	139 La* lanthanum 57	178 Hf hafnium 72	181 Ta tantalum 73	184 W tungsten 74	186 Re rhenium 75	190 Os osmium 76	192 Ir iridium 77	195 Pt platinum 78	197 Au gold 79	201 Hg mercury 80		204 Tl thallium 81	207 Pb lead 82	209 Bi bismuth 83	[209] Po polonium 84	[210] At astatine 85	[222] Rn radon 86
[223] Fr francium 87	[226] Ra radium 88	[227] Ac* actinium 89	[261] Rf rutherfordium 104	[262] Db dubnium 105	[266] Sg seaborgium 106	[264] Bh bohrium 107	[277] Hs hassium 108	[268] Mt meitnerium 109	[271] Ds darmstadtium 110	[272] Rg roentgenium 111								

Elements with atomic numbers 112–116 have been reported but not fully authenticated.

* The Lanthanides (atomic numbers 58–71) and the Actinides (atomic numbers 90–103) have been omitted.

Cu and Cl have not been rounded to the nearest whole number.

Exam tips

The key to successful revision is finding the method that suits you best. There is no right or wrong way to do it.

Before you begin, it is important to plan your revision carefully. If you have allocated enough time in advance, you can walk into the exam with confidence, knowing that you are fully prepared.

Start well before the date of the exam, not the day before!

It is worth preparing a revision timetable and trying to stick to it. Use it during the lead up to the exams and between each exam. Make sure you plan some time off too.

Different people revise in different ways, and you will soon discover what works best for you.

Remember!

There is a difference between *learning* and *revising*.

When you revise, you are looking again at something you have already learned. Revising is a process that helps you to remember this information more clearly.

Learning is about finding out and understanding new information.

Some general points to think about when revising

- Find a quiet and comfortable space at home where you won't be disturbed. You will find you achieve more if the room is ventilated and has plenty of light.

- Take regular breaks. Some evidence suggests that revision is most effective when tackled in 30 to 40 minute slots. If you get bogged down at any point, take a break and go back to it later when you are feeling fresh. Try not to revise when you're feeling tired. If you do feel tired, take a break.

- Use your school notes, textbook and this Revision guide.

- Spend some time working through past papers to familiarise yourself with the exam format.

- Produce your own summaries of each module and then look at the summaries in this Revision guide at the end of each module.

- Draw mind maps covering the key information on each topic or module.

- Review the **Grade booster checklists** on page 152–159.

- Set up revision cards containing condensed versions of your notes.

- Prioritise your revision of topics. You may want to leave more time to revise the topics you find most difficult.

Workbook

The **Workbook** (pages 87–151) allows you to work at your own pace on some typical exam-style questions. These are graded to show the level you are working to (G–E, D–C or B–A*). You will find that the actual GCSE questions are more likely to test knowledge and understanding across topics. However, the aim of the Revision guide and Workbook is to guide you through each topic so that you can identify your areas of strength and weakness.

The Workbook also contains example questions that require longer answers (**Extended response questions**). You will find one question that is similar to these in each section of your written exam papers. The quality of your written communication will be assessed when you answer these questions in the exam, so practise writing longer answers, using sentences. The **Answers** to all the questions in the Workbook can be cut out for flexible practice and can be found on pages 160–168.

Collins Workbook

NEW GCSE

Chemistry

OCR

Twenty First Century Science

Authors: Brian Cowie
Ann Tiernan

Revision Guide +
Exam Practice Workbook

The changing air around us

1 a Underline the **two** main gases that make up the air.

helium   hydrogen   oxygen   nitrogen   **[2 marks]**

b Finish the sentence by choosing the best word from this list:

compound   element   mixture   pressure

Air in the atmosphere is a gas ....   **[1 mark]**

c Why are clouds not parts of the air?

....   **[1 mark]**

G–E

2 This diagram represents a molecule of oxygen. ◯◯

This diagram represents a molecule of nitrogen. ●●

Use the diagrams to draw a representation of a sample of air in the box below.

D–C

[2 marks]

3 Describe an experiment that can be used to find the percentage of oxygen in the air.

....

....

....

....   **[3 marks]**

B–A*

4 a Tick **two** boxes showing the gases released by volcanoes.

carbon dioxide   ☐

chlorine   ☐

hydrogen   ☐

water vapour   ☐   **[2 marks]**

b What did these two gases make about 4 billion years ago?

....   **[1 mark]**

G–E

5 These sentences describe how our present atmosphere formed. They are not in the correct order.

i Plants produced oxygen.

ii Simple bacteria removed carbon dioxide by photosynthesis.

iii The Earth cooled, allowing water vapour to condense.

iv Four billion years ago the atmosphere was very hot.

v Fossil fuels formed from buried organisms.

The correct order of the sentences is: ....   **[4 marks]**

D–C

6 Most scientists now agree on the explanation of how our atmosphere formed.

What is needed for a theory to become accepted?

....

....

....

....   **[3 marks]**

B–A*

Humans, air quality and health

1 a Name **two** uses of fuels.

.... **[2 marks]**

b Tick **one** box that shows a gas that is not an air pollutant.

carbon monoxide ☐

carbon dioxide ☐

hydrogen oxide ☐

sulfur dioxide ☐

[1 mark]

c What is meant by 'good' air quality?

.... **[1 mark]**

2 This chart shows how carbon dioxide levels have increased over the last 50 years.

Year	1960	1970	1980	1990	2000	2010
Carbon dioxide levels (ppm)	310	325	340	350	370	395

a Describe the pattern shown by the data.

....

....

....

.... **[3 marks]**

b Suggest how solid particulates get into the air by:

i a natural process: .... **[1 mark]**

ii human activity: .... **[1 mark]**

c Name the **two** pollutants that contribute to acid rain.

.... **[2 marks]**

d Name **two** health problems made worse by poor air quality.

.... **[2 marks]**

3 a The table in Question 2 shows units of ppm. What is ppm and what does it mean?

....

.... **[2 marks]**

b In the UK air quality is regularly monitored. Suggest why.

....

....

....

.... **[3 marks]**

4 Airborne carbon particulates have been linked to lung disease. Lung disease is also linked to smoking.

a Describe why the facts in both sentences are correlations.

....

.... **[1 mark]**

b Explain why a correlation between smoking and lung disease does not necessarily mean this factor is the cause.

....

.... **[2 marks]**

G–E

D–C

B–A*

B–A*

Burning fuels

1 a Name the **two** elements in hydrocarbons. .... [2 marks]

   b When coal burns, carbon joins with oxygen to form carbon dioxide.

     Use the information to write the word equation for this reaction.

     .... [2 marks]

G–E

2 a Complete this word equation for combustion:

     hydrocarbon fuel + ... $\longrightarrow$ .... + ... [2 marks]

   b Underline a word from the list below to describe this reaction: sulfur + oxygen $\longrightarrow$ sulfur dioxide.

     acidification   combustion   neutralisation   oxidation   reduction   [1 mark]

D–C

3 Gas welding uses ethyne gas (C_2H_2) and oxygen (O_2).

   Draw a visual representation to show this reaction.

   .... [3 marks]

B–A*

4 Use the best words from this list to complete the sentences.

   joined   complex   molecules   new   old   rearranged

   a Atoms of non metal elements join to make .... [1 mark]

   b In chemical reactions, atoms are .... to make

     .... substances. [2 marks]

G–E

5 Complete these sentences.

   a The substances on the left side of a chemical equation are the ....

     and the right side shows the .... [2 marks]

   b The total mass on each side of a chemical equation is .... [1 mark]

D–C

6 All the atoms in all the fossil fuels ever burnt are still present on Earth. Explain why.

   ....

   .... [2 marks]

B–A*

7 a Burning sulfur in air is a chemical reaction. State two things you see that shows this.

     .... [2 marks]

   b Give two differences between sulfur and sulfur dioxide.

     .... [2 marks]

G–E

8 a Explain why fossil fuels contain sulfur.

     ....

     .... [2 marks]

   b Write a symbol equation for burning sulfur.

     .... [2 marks]

D–C

9 a Describe how acid rain forms, and how it damages the environment.

     ....

     ....

     .... [3 marks]

   b Why is acid rain described as a indirect pollutant?

     .... [1 mark]

B–A*

Pollution

1 a Name two processes that burn large amounts of fuel.

.... [2 marks]

b When carbon burns in a good air supply, carbon dioxide is formed. What is made if carbon is

burnt in a limited oxygen supply? .... [1 mark]

c Which element in fuels becomes sulfur dioxide when the fuel burns? .... [1 mark]

2 Name the two gases in the air responsible for nitrogen oxide formation.

.... [2 marks]

G–E

3 a During high temperature combustion, nitrogen monoxide is released.
Draw a visual representation to show this reaction.

.... [3 marks]

b Describe, giving examples, why nitrogen oxides are classified as pollutants.

....

....

.... [3 marks]

B–A*

4 Name these compounds.

a .... **b** .... **c** .... [3 marks]

D–C

5 Suggest natural processes that remove the following pollutants from the air.

a Particulate carbon: .... [1 mark]

b Sulfur and nitrogen oxides: .... [1 mark]

c Carbon dioxide: .... [1 mark]

6 This data shows the levels of nitrogen oxides in air on one street.

All the data was taken over a 5-minute period.

Reading	1	2	3	4	5	6
Value (units)	15	12	14	8	14	12

a Suggest a reading that is likely to be an outlier. .... [1 mark]

b Write down the range of the results. .... [1 mark]

c Calculate the mean value. Show your working out.

....

.... [2 marks]

D–C

7 Air quality measurements occasionally produce unexpected results.

Explain why unexpected results are not necessarily outliers.

....

....

.... [2 marks]

B–A*

Improving power stations and transport

1 Over the last decade, many new gas-fired power stations have been built.

Suggest **two** advantages of producing electricity from gas rather than coal.

....

.... [2 marks]

D–C

2 Suggest **two** methods of removing sulfur dioxide at power stations.

....

.... [2 marks]

B–A*

3 Suggest **two** ways of saving electricity in the home.

....

.... [2 marks]

G–E

4 Why is it important to save fossil fuels?

.... [1 mark]

5 Some power stations burn biofuels. Give **two** examples of biofuels.

.... [2 marks]

D–C

6 a Explain how biofuels are made, and why they are classed as 'carbon neutral'.

....

.... [3 marks]

b Describe **two** disadvantages of replacing fossil fuels with biofuels.

B–A*

....

.... [2 marks]

7 Cars and lorries release pollutants into the air.

Suggest **two** ways the amount of pollutants could be reduced.

G–E

.... [2 marks]

8 Catalytic convertors contain a platinum catalyst than helps pollutant gases react.

a Where are catalytic convertors found in cars?

.... [1 mark]

b Complete this word equation to show how the pollutant gases are removed.

Carbon monoxide + ... ⟶ .... + ... [2 marks]

D–C

c Describe why the above reaction is both oxidation and reduction.

....

.... [2 marks]

9 Many motor companies are developing electric cars.

a Suggest problems of using electric cars at present.

....

.... [3 marks]

B–A*

b Explain why using electric cars is not 'carbon neutral'.

....

.... [2 marks]

Air quality varies from day to day.

This chart shows the carbon particulate level (PM10) and nitrogen oxide levels (NO$_x$) in Birmingham City Centre over a week in January.

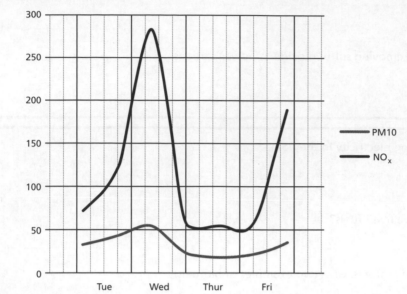

Describe what the graph shows. Include ideas about how the pollutants change, whether a correlation exists, and possible causes.

✒ *The quality of written communication will be assessed in your answer to this question.*

B–A*

[6 marks]

Using and choosing materials

1 a Draw lines to match each material to its best property and to its use.

Material	Property	Use
Rubber	Can be moulded	Washing-up bowls
Plastic	Hard and elastic	Making clothes
Fibres	Can be woven	Car tyres

[2 marks]

b What does the word 'property' mean when talking about materials?

....

.... [1 mark]

2 Write in the missing property to complete each sentence.

a The temperature at which a solid turns to a liquid: .... [1 mark]

b The force needed to crush a material: .... [1 mark]

c How well a material stands up to wear: .... [1 mark]

d The mass of a given volume of material: .... [1 mark]

3 Climbers need to consider a number of properties when buying a rope.

a Suggest why climbing ropes need to be:

i dynamic (stretchy)

.... [1 mark]

ii dry-treated (waterproof)

.... [1 mark]

b Single ropes need to be a minimum of 9 mm. Why are 11-mm ropes stronger?

....

.... [2 marks]

4 Bill and Joy measured how far a fibre stretched when 200 g was added. Here are the results.

Result	1	2	3	4	5	6
Stretch (cm)	5.8	5.7	5.3	5.6	5.8	5.6

a Identify the outlier in the results. .... [1 mark]

b State the range. .... [1 mark]

c Why do many results need to be taken?

.... [1 mark]

d What is the true value of these results? .... [1 mark]

e Suggest **two** reasons for an outlier being discarded.

....

....

.... [2 marks]

G–E

D–C

B–A*

D–C

B–A*

Natural and synthetic materials

1 Underline **three** properties which most metals have.

   brittle   good conductor   hard   insulator   malleable   soft   [3 marks]

2 Give **three** examples of ceramics.

   ....

   .... [3 marks]

3 Rubber, plastic and fibres are all made of large molecules called .... [1 mark]

4 Give one example of a natural material which comes from:

   a plants ....

   b animals ....

   c the Earth's crust .... [3 marks]

5 What are synthetic materials?

   ....

   ....

   ....

   .... [2 marks]

6 Give **three** reasons why synthetic materials have replaced some natural materials.

   ....

   ....

   .... [3 marks]

7 Name the two elements present in hydrocarbon molecules.

   .... [2 marks]

8 Finish the sentences.

   a Most crude oil is used for making .... [1 mark]

   b When oil burns the number of atoms that react are .... to
   the number of atoms of products made. [1 mark]

9 The general formula for an alkane is C_nH_{2n+2}

   a What would be the formula for butane, which contains 4 carbon atoms?

   .... [1 mark]

   b What would be the formula for hexane, which contains 6 carbon atoms?

   .... [1 mark]

10 Suggest why the composition of crude oil varies from place to place.

   ....

   ....

   .... [1 mark]

11 When hydrocarbon fuels burn in oxygen, carbon dioxide and water are made.
   One hydrocarbon fuel is methane, CH_4.

   Draw a visual representation to show the complete combustion of methane.

   .... [3 marks]

G–E

D–C

B–A*

G–E

D–C

B–A*

Separating and using crude oil

1 Finish the sentences. Choose the best words from this list:

condenses   distillation   evaporates   fractions   molecules

a Crude oil is separated by fractional .... **[1 mark]**

b The oil is heated up so it .... **[1 mark]**

c When the gas rises up and cools it .... into a liquid. **[1 mark]**

d Gases with similar boiling points collect together into .... **[1 mark]**

2 Describe the link between the size of hydrocarbon molecules and the boiling point.

.... **[1 mark]**

3 Explain why petrol boils at a lower temperature than diesel.

....

....

....

.... **[3 marks]**

4 a Draw and label paper clips to show the difference between monomers and polymers.

[1 mark]

b Polymers have replaced many natural materials. Name a polymer that has replaced:

i aluminium for tennis rackets: .... **[1 mark]**

ii metal for buckets: .... **[1 mark]**

5 The expanded formula for the monomer ethene is shown below.

Next to it, draw the expanded formula for polyethene.

$$\begin{array}{c} H \qquad H \\ \diagdown \qquad \diagup \\ C = C \\ \diagup \qquad \diagdown \\ H \qquad H \end{array}$$

[2 marks]

6 Describe why PET polymer is superior to glass for making bottles.

....

....

.... **[3 marks]**

7 Describe how the properties of polymer chains can be changed.

....

.... **[2 marks]**

Polymers: properties and improvements

G–E

1 Look at these polymer molecules.

i

ii

iii

Which molecule will have:

a the strongest forces between molecules? .... [1 mark]

b the highest melting point? .... [1 mark]

D–C

2 Low density polyethene (LDPE) has long molecules with many branches.

a Suggest what the properties of LDPE are likely to be.

....

....

.... [3 marks]

b Describe the structure of HDPE and why it is likely to have a higher melting point than LDPE.

....

....

.... [3 marks]

B–A*

3 Explain how the degree of crystallinity affects a polymer.

....

....

....

.... [3 marks]

G–E

4 Finish the sentences. Choose the best words from this list:

higher   less   lower   more   stronger   weaker

a The longer the chain length, the .... the polymer.   [1 mark]

b Longer chains need .... force to separate them.   [1 mark]

c Longer chains have a .... melting point.   [1 mark]

D–C

5 Describe plasticisers and how they work.

....

....

....

.... [4 marks]

6 How are thermoplastics different to thermosetting plastics?

....

....

.... [3 marks]

7 Natural rubber is too soft for using as a car tyre. Suggest **two** ways the rubber can be made harder.

....

.... [2 marks]

B–A*

8 a Describe how crystallinity can be increased.

....

.... [1 mark]

b Suggest a disadvantage of increased crystallinity.

....

.... [1 mark]

Nanotechnology and nanoparticles

1 Underline the instrument used to view very small objects.

bathyscope   milliscope   microscope   telescope   [1 mark]

2 The width of a human hair is about 0.1 mm. How many would fit into 1 mm?

....   [1 mark]

3 a Name **one** naturally occurring nanoparticle.

....   [1 mark]

  b What is the name of the nanoparticles released when a fuel burns?

....   [1 mark]

  c Nanoparticles can occur naturally, or by accident. Name another way they occur.

....   [1 mark]

G–E

4 Underline the correct answer. Nanoparticles are about the same size as a:

human hair   large molecule   oxygen molecule   salt crystal   [1 mark]

5 How many nanometres are in a metre? ....   [1 mark]

6 Explain the difference between 'buckyballs' and nanotubes.

Buckyball   Nanotubes

....

....

....

....   [4 marks]

7 Explain why nanoparticles are effective catalysts.

....

....   [2 marks]

D–C

8 a Explain why a carbon nanotube can have a diameter of just a few nanometres but a length of many millimetres.

....

....   [2 marks]

  b Explain why cutting a 1 cm³ cube into four means the volume stays the same but the surface area increases.

....

....

....

....   [4 marks]

B–A*

The use and safety of nanoparticles

G–E

1 Suggest why nanoparticles are added to socks.

.... [1 mark]

2 Nanoparticles can be added to tennis balls to make them bouncy for longer. Look at these results.

Test / date	Bounce height in cm from 100 cm height			
	New	3 months old	6 months old	9 months old
1	79	75	78	77
2	81	80	77	76
3	82	79	78	77
4	80	79	79	76

D–C

a Calculate the mean result when the ball is new. .... [1 mark]

b State the range shown by the 6-months-old results. .... [1 mark]

c Identify one outlier contained in the data. .... [1 mark]

d Describe if the data shows whether or not the ball has stayed bouncy.

....

....

.... [3 marks]

e What name is given to a mixture of materials? .... [1 mark]

3 What is meant by the term nanotechnology?

.... [1 mark]

4 Carbon nanotubes are being incorporated into body armour. Suggest why this makes the armour strong, low density and flexible.

B–A*

....

....

....

.... [3 marks]

G–E

5 Silver nanoparticles can be washed out of clothing. Why can this be a problem?

....

.... [2 marks]

D–C

6 Nanoparticles can be incorporated into sunscreen to prevent damage by ultraviolet rays. Suggest why some people are concerned about the safety of this type of sunscreen.

....

....

.... [3 marks]

B–A*

7 Some environmental and health groups are campaigning to have nanoparticles banned. Outline arguments **against** this point of view.

....

....

....

.... [3 marks]

C2 Extended response question

Badminton rackets can be made using different materials.

The table shows information about badminton racket design for players of different abilities.

Rackets for...	Stem material	Mass (g)	Cost (£)
Beginner	Titanium and steel	120	15
Intermediate	Graphite and titanium	95	30
Advanced	Graphite	80	45
Elite	Nano-filled graphite	85	70

With reference to the table, describe the properties that are needed in a badminton racket, and explain why the materials used result in these different properties.

✎ The quality of written communication will be assessed in your answer to this question.

[6 marks]

Moving continents and useful rocks

1 Underline the name that describes scientists who study rocks.

astronomers   geologists   meteorologists   oceanologists   **[1 mark]**

2 Tectonic plates can move by sliding past each other. State **two** other ways they can move.

.... **[2 marks]**

3 A supercontinent called Pangea existed about 225 million years ago. Describe how this happened, and what formed as a result.

....

.... **[2 marks]**

4 Why has the climate in Britain changed over the last 600 million years?

....

.... **[2 marks]**

5 Describe how magnetism can help date igneous rock.

....

....

....

.... **[4 marks]**

6 a Why did the Industrial Revolution start in the north-west of Britain?

.... **[1 mark]**

b Name **three** raw materials needed for the Industrial Revolution.

.... **[3 marks]**

7 Describe the conditions needed for limestone to form.

....

.... **[3 marks]**

8 Describe how rock salt deposits form.

....

....

.... **[4 marks]**

9 a What evidence does this picture show about this sedimentary rock?

....

....

....

.... **[2 marks]**

b Describe **one** other form of evidence which can be used to show the conditions needed for a sedimentary rock.

....

.... **[2 marks]**

Salt

1 Name **two** different ways used to obtain salt.

.... [2 marks]

2 State why salt is put onto icy roads and explain how it works.

....

....

.... [3 marks]

3 Suggest why salt is not obtained from sea water in the UK.

.... [1 mark]

4 Describe **two** problems with salt extraction and their environmental impact.

....

....

....

....

....

.... [4 marks]

5 Give **two** uses of salt in the food industry.

....

.... [2 marks]

6 The table shows how salt affects bacterial growth.

% salt concentration	0	12	24	36	48	60
Number of bacteria	58	54	50	46	32	5

Describe the pattern shown by the data.

....

.... [3 marks]

7 a Explain why salt is classified as a possible hazard, and how the risks can be estimated.

....

....

....

.... [3 marks]

b Describe how people can reduce the risk from salt.

....

.... [2 marks]

8 Describe how the UK government regulates food safety.

....

....

....

....

.... [4 marks]

Reacting and making alkalis

1 Tick **two** boxes showing properties of alkalis.

   a pH of between 1 and 6 ☐   **c** Turn litmus paper blue ☐

   b Convert stale urine to potassium hydroxide ☐   **d** Neutralise acid soil and are used to make glass ☐

                                                [2 marks]

2 Alkalis react with acids to make salts. If nitric acid reacts with calcium hydroxide, a salt called calcium nitrate is made. What salt is made when:

   a sulfuric acid reacts with potassium hydroxide? .... **[1 mark]**

   b hydrochloric acid reacts with sodium hydroxide? .... **[1 mark]**

3 Name **two** traditional sources of alkalis.

.... **[2 marks]**

4 Before industrialisation, alkalis were in short supply. Explain why.

....

....

.... **[3 marks]**

5 Lime is made by heating limestone. Limestone has the formula $CaCO_3$. Carbon dioxide gas (CO_2) is also produced.

   a Draw a visual representation to show the reaction for making lime.

.... **[1 mark]**

   b Lime is used to neutralise acidic soil. Describe why this is preferable to adding an alkali.

....

....

....

....

.... **[4 marks]**

6 Describe the role of a mordant in the dying process.

.... **[1 mark]**

7 The Leblanc process manufactured sodium carbonate by reacting limestone and salt.

   a Name two toxic by-products of this process.

.... **[2 marks]**

   b Explain how the pollution problems arising from this process can be solved.

....

.... **[2 marks]**

8 a Predict the product of this reaction:

     sodium carbonate + sulfuric acid ⟶ .... **[1 mark]**

   b Predict the product of this reaction:

     ammonium hydroxide + nitric acid ⟶ .... **[1 mark]**

9 All metal hydroxides are classed as bases. Why are some metal hydroxides alkalis?

....

.... **[2 marks]**

C3 Chemicals in our lives – risks and benefits

Uses of chlorine and its electrolysis

1 This bar chart shows deaths from typhoid in different countries.

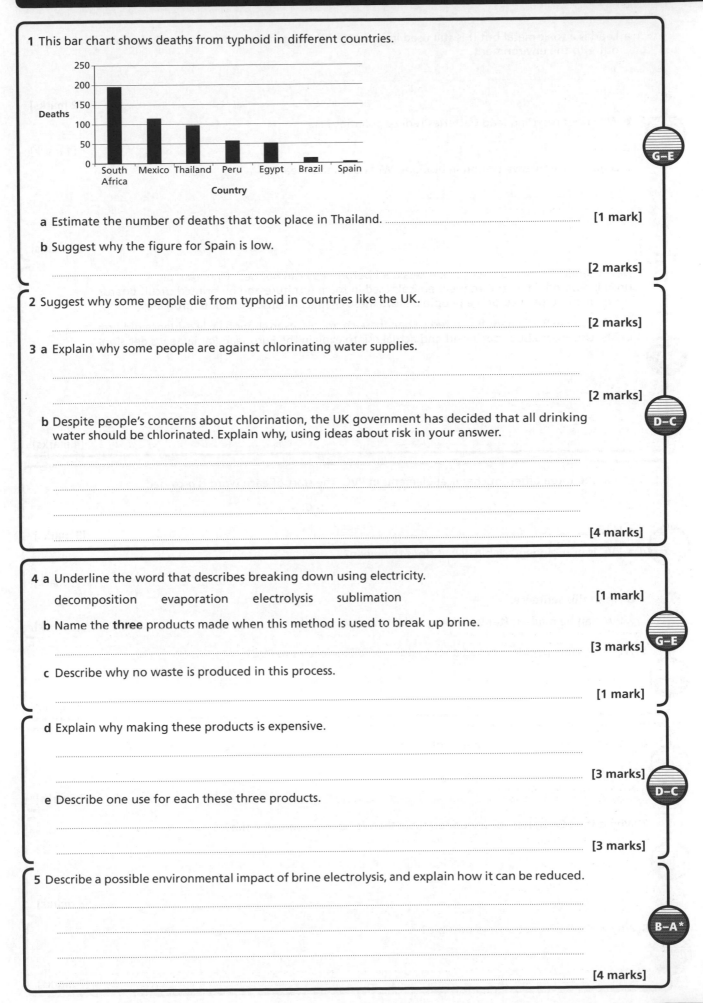

a Estimate the number of deaths that took place in Thailand. .... [1 mark]

b Suggest why the figure for Spain is low.

.... [2 marks]

2 Suggest why some people die from typhoid in countries like the UK.

.... [2 marks]

3 a Explain why some people are against chlorinating water supplies.

....

.... [2 marks]

b Despite people's concerns about chlorination, the UK government has decided that all drinking water should be chlorinated. Explain why, using ideas about risk in your answer.

....

....

....

....

.... [4 marks]

4 a Underline the word that describes breaking down using electricity.

decomposition   evaporation   electrolysis   sublimation   [1 mark]

b Name the **three** products made when this method is used to break up brine.

.... [3 marks]

c Describe why no waste is produced in this process.

.... [1 mark]

d Explain why making these products is expensive.

....

.... [3 marks]

e Describe one use for each these three products.

....

.... [3 marks]

5 Describe a possible environmental impact of brine electrolysis, and explain how it can be reduced.

....

....

....

.... [4 marks]

Industrial chemicals and LCA

1 a Lead is a toxic metal but it is still used in car batteries. Suggest **two** ways lead could spread out into the environment.

....

.... [2 marks]

b Why does recycling lead batteries reduce pollution?

.... [1 mark]

c Lead is a cumulative poison in humans. What does this mean?

....

....

.... [3 marks]

2 New chemicals need to pass a risk assessment before they can be used. One example is bromine-based flame-retardant chemicals used in foam furniture and on printed circuit boards for games and phones. Some people believe these chemicals cause behaviour problems.

The European Parliament has voted to ban these chemicals, even though little evidence of harm exists. Use ideas about perceived and actual risk to explain the European Parliament's decision.

....

....

....

.... [4 marks]

3 Name the **three** different chemical elements in PVC. The start of each word is provided.

a C is .... **b** Cl is ....

c H is .... [3 marks]

4 a PVC is non-biodegradable. What does this mean?

.... [1 mark]

b Finish this sentence:

PVC can be made softer by adding small molecules called .... [1 mark]

5 Why are some people worried about adding small molecules to PVC?

....

.... [2 marks]

6 Explain why a Life Cycle Assessment is sometimes referred to as 'cradle to grave' assessment.

....

.... [3 marks]

7 Why is the LCA better for a wooden table than for using wood on a fire?

....

....

.... [3 marks]

8 Why is a full LCA not always possible?

....

.... [2 marks]

C3 Chemicals in our lives – risks and benefits

C3 Extended response question

Many local councils collect plastic bottles for recycling into new plastics. Plastics can also be made directly from crude oil.

The two flow charts show the different processes.

Collecting and recycling

Collection by lorry from homes

↓

Sort into types and clean

↓

Crush and shred

↓

Make into pellets

Making from crude oil

Drill for oil and distil it

↓

'Crack' chains to make monomers

↓

Join monomers to make polymers

↓

Make into pellets

Describe what is meant by the term Life Cycle Assessment, and use the charts to compare the Life Cycle Assessment of the two methods.

🖉 *The quality of written communication will be assessed in your answer to this question.*

B–A*

[6 marks]

Atoms, elements and the Periodic Table

1 Mendeleev used data about the elements to arrange them into the Periodic Table.

Give **two** types of data that Mendeleev used.

In their properties and relative atomic mass

[2 marks]

2 The table shows three elements that Döbereiner put into a 'triad'.

Elements	Relative atomic mass
Top element: lithium	7
Middle element: sodium	23
Bottom element: potassium	39

a Work out the mean of the relative atomic mass of the top and bottom element.

23 mean relative atomic mass

[1 mark]

b Explain why Döbereiner thought sodium fitted as the middle element.

because it has an average relative atomic mass

[1 mark]

3 Other scientists, such as Newlands and Mendeleev, had different ideas about how to organise the elements.

a Both Newlands and Mendeleev also put lithium, sodium and potassium together in their arrangement of the elements.

Explain why they both thought that these three elements belong together.

because they all had similar properties

[1 mark]

b Give **two** reasons why Mendeleev's arrangement of elements was an improvement on Newlands' arrangement.

[2 marks]

4 The diagram shows the line spectrum of helium and hydrogen.

Scientists used line spectra of elements to find out that hydrogen and helium are in our Sun.

Explain how they did this.

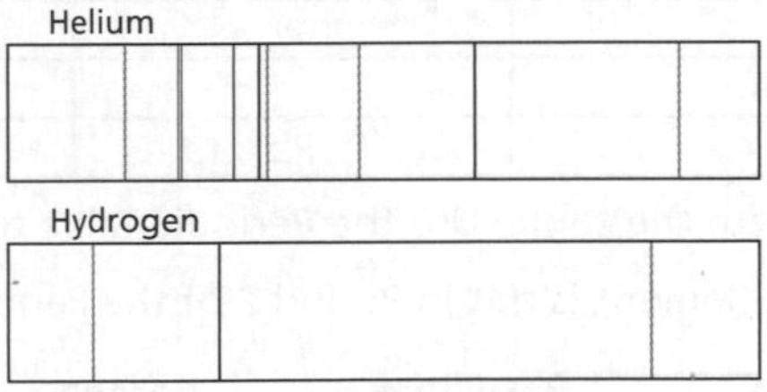

Helium

Hydrogen

[3 marks]

5 Complete the sentences about the structure of an atom. Choose words from the list below.

electrons   groups   ions   molecules   neutrons   protons   shells

An atom contains a tiny nucleus, which contains particles called ....

and .... travel around the outside

of the atom. They are arranged in ....

[2 marks]

6 The table shows information about some atoms. Complete the table.

Proton number	Relative atomic mass	No. of protons	No. of neutrons	No. of electrons
9	19			
	27			13
			4	3

[3 marks]

Electrons and the Periodic Table

1 Lithium and sodium are both elements in Group 1 of the Periodic Table.

The table shows some information about a lithium and a sodium atom.

	No. of protons	No. of neutrons	No. of electrons
lithium	3	4	3
sodium	11	12	11

a The diagram below left shows the electron arrangement in a lithium atom.

Complete the diagram and labels to show the electron arrangement in a sodium atom.

Lithium

Sodium

Electron arrangement: 2.1   Electron arrangement: .... **[2 marks]**

b Potassium is another element in Group 1. Potassium has a proton number of 19.

Describe the **similarities** and **differences** between the electron arrangements of potassium and sodium.

....

....

.... **[3 marks]**

2 a The table shows some information about some elements.

Name of element	Symbol	Proton number
neon		10
fluorine		
	Pb	82

Complete the table. Use the Periodic Table to help you. **[3 marks]**

b Which element is **not** in Period 2 of the Periodic Table? Put a ring around your choice.

Lithium   **Beryllium**   **Carbon**   **Aluminium**   **Oxygen**   **[1 mark]**

3 The following statements are about trends across a period of the Periodic Table. Put a tick (✓) in the boxes next to the **two** correct statements.

From left to right across a period in the Periodic Table...

a the number of electrons in the outer shell increases. ☐

b the elements are more likely to be non-metals. ☐

c proton number decreases. ☐

d the elements are more likely to be solids. ☐   **[2 marks]**

4 A textbook gives this statement: *The number of electrons in the outer shell gives information about the group number and the metal or non-metal character of the element.*

Explain what this statement means.

....

....

....

.... **[3 marks]**

C4 Chemical patterns

Reactions of Group 1

1 The table shows some data about Group 1 elements.

Element	Melting point in °C	Boiling point in °C
lithium	180.0	1330
sodium	97.8	
potassium	63.7	774

a What is the trend in melting point down Group 1?

.... [1 mark]

b Predict the boiling point of sodium. Explain your reasoning.

.... [1 mark]

2 a Joe adds a piece of sodium to some water that contains pH indicator. The boxes show his observations during the reaction.

Use straight lines to connect each observation with each reason.

Observation

a The sodium goes into a ball shape.	

b The sodium fizzes and bubbles form.	

c The pH indicator turns blue.	

Reason

i Hydrogen is made.	

ii Sodium hydroxide is made.	

iii The metal melts because the reaction gives out energy.	

[2 marks]

b Joe repeats the experiment. This time he uses potassium rather than sodium. Give **two** similarities and **two** differences between the reactions of potassium and sodium with water.

....

....

....

[4 marks]

3 Eve investigates the reactions of Group 1 metals with chlorine.

a She reacts a hot piece of lithium with chlorine.

Complete and balance the symbol equation for the reaction.

Li   + $\longrightarrow$ LiCl   [2 marks]

b Eve reacts sodium and potassium with chlorine. She notices that the reactions get faster when she uses a metal further down the group.

She writes this conclusion in her notes: *I think that for Group 1 elements, the more electron shells there are in the atom, the more reactive the element.*

Do you agree with Eve? Use ideas about electron arrangement to explain your reasoning.

....

....

....

[4 marks]

Group 7 – The halogens

G–E

1 Draw straight lines to connect each **halogen** to its **state** and **colour** at room temperature and pressure.

State	Halogen	Colour
		red-brown
solid	chlorine	
		dark grey
liquid	bromine	
		purple
gas	iodine	
		pale green

[3 marks]

D–C

2 The halogens all contain **diatomic** molecules.

Give the formula for a chlorine molecule and explain why it is diatomic.

.... [2 marks]

3 The states of the halogens show a trend down the group.

a Describe the trend in the states of the halogens down the group.

.... [1 mark]

b Describe **one** other trend shown by the halogens down the group.

....

.... [2 marks]

G–E

4 Rose heats some iron wool and puts it into a gas jar that contains chlorine.

The iron wool glows very hot and a brown solid is formed.

a What is the name of the brown solid that is made in the reaction?

b What safety precautions should Rose take when she uses chlorine?

Explain your reasoning.

....

.... [2 marks]

D–C

c Rose does the experiment again. This time she reacts iron with bromine gas.

How does the rate of the reaction change when Rose uses bromine instead of chlorine? Explain your answer.

....

.... [2 marks]

B–A*

5 Astatine is a halogen. It is below iodine in Group 7.

a Jane adds some chlorine water to a solution of potassium astatide.

Complete and balance the symbol equation for the reaction that happens.

$$Cl_2 \quad + \quad KAt \quad \longrightarrow \quad KCl \quad + \qquad \text{[2 marks]}$$

b Chlorine is much more reactive than astatine. Use ideas about electron shells to explain why.

....

.... [2 marks]

Ionic compounds

1 Which of the statements about ionic compounds are true? Put a tick (✓) in the boxes next to the **two** correct answers.

a Ionic compounds are usually gases. ☐

b Ions in a solid ionic compound are arranged in a regular pattern. ☐

c Compounds of Group 1 and Group 7 elements are ionic. ☐

d Ionic compounds have low melting points. ☐

e Ionic compounds conduct electricity when they dissolve in water. ☐

[2 marks]

G–E

2 The table shows information about some chemicals.

Chemical	Does it conduct electricity when solid?	Does it conduct electricity when melted?	Melting point
A	no	yes	high
B	no	no	low
C	yes	yes	high

a Which chemical, A, B or C, is a metal?

.... **[1 mark]**

b Which chemical, A, B or C, is an ionic compound?

.... **[1 mark]**

G–E

3 a How does the electrical conductivity of a solid ionic compound change when its temperature increases above its melting point?

....

.... **[2 marks]**

b Use ideas about ions to explain why the electrical conductivity changes.

....

.... **[2 marks]**

D–C

4 The table gives some information about a sodium atom and a chlorine atom.

	No. of protons	No. of neutrons	No. of electrons
sodium atom	11	12	11
chlorine atom	9	10	9

The sodium atom reacts to become a sodium ion.

a Give **two** similarities and **one** difference between a sodium atom and a sodium ion. Use the table to help you.

....

.... **[3 marks]**

b Give **one** difference between what happens when a sodium atom becomes an ion and when a chlorine atom becomes an ion.

.... **[2 marks]**

D–C

5 The formula for sodium chloride is NaCl. The formula for calcium chloride is $CaCl_2$.

Use ideas about charges on the ions to explain why the formulae are different.

....

....

....

.... **[4 marks]**

B–A*

C4 Extended response question

Lithium is at the top of Group 1 in the Periodic Table. Fluorine is at the top of Group 7 in the Periodic Table.

The position of an element in the Periodic Table gives information about:

- the arrangement of the outer-shell electrons.
- whether the element is a metal or a non-metal
- the reactivity of the element.

Describe the differences between lithium and fluorine based on their positions in the Periodic Table.

✎ *The quality of written communication will be assessed in your answer to this question.*

[6 marks]

Molecules in the air

1 Air contains a mixture of gases

Draw straight lines to connect each **gas** with its correct **formula** and **percentage** in the air.

Formula	Gas	Percentage
O_2	nitrogen	21%
Ar	oxygen	78%
CO_2	argon	0.04%
N_2	carbon dioxide	about 1%

[2 marks]

G–E

2 The atoms in the molecules of gases in the air are held together by covalent bonds.

Which of the following statements about covalent bonds are true? Put a tick (✓) in the boxes next to the **two** correct answers.

a When a covalent bond forms, atoms lose or gain electrons. ☐

b The nuclei of both atoms are attracted to the electrons in the bond. ☐

c Atoms share electrons to form a bond. ☐

d Very little energy is needed to overcome the attraction between atoms. ☐

e The bonds are always arranged in a 2-D arrangement. ☐

[2 marks]

B–A*

3 The table shows data about some elements in the air.

Substance	Melting point (°C)	Boiling point (°C)
nitrogen	−210	−196
oxygen	−218	−183
argon	−189	−186

a Which element is a liquid over the largest temperature range?

.... **[1 mark]**

b Use ideas about forces between molecules to explain why the melting points and boiling points of the elements are all very low.

....

.... **[2 marks]**

D–C

c Ben looks at the table. He writes this note in his book: *Nitrogen has the lowest boiling point. There is a link because gases with very low melting points have very low boiling points.*

Do you agree with Ben? Explain your reasoning.

....

....

....

....

.... **[4 marks]**

B–A*

1 The table shows some information about solid sodium chloride.

Melting point	Solubility in water	Electrical conductivity
801 °C	very soluble	does not conduct when solid

a Explain why solid sodium chloride has a high melting point.

....

.... **[3 marks]**

b Explain why sodium chloride does not conduct electricity when it is solid.

.... **[1 mark]**

c Describe what happens to the arrangement and movement of the ions when sodium chloride dissolves in water.

....

....

.... **[2 marks]**

2 The table shows the ions and formulae for some ionic compounds. Complete the table.

Name	Positive ion	Negative ion	Formula
potassium chloride	K^+	Cl^-	KCl
calcium chloride	Ca^{2+}	Cl^-	
potassium sulfate			
calcium sulfate	Ca^{2+}	SO_4^{2-}	$CaSO_4$

[3 marks]

3 The table below shows the results for tests in which dilute sodium hydroxide was added to known cations.

Ion	Observation
calcium Ca^{2+}	a white precipitate forms; the precipitate does not dissolve in excess sodium hydroxide
copper Cu^{2+}	a light blue precipitate forms; the precipitate does not dissolve in excess sodium hydroxide
iron(II) Fe^{2+}	a green precipitate forms; the precipitate does not dissolve in excess sodium hydroxide
iron(III) Fe^{3+}	a red-brown precipitate forms; the precipitate does not dissolve in excess sodium hydroxide
zinc Zn^{2+}	a white precipitate forms; the precipitate dissolves in excess sodium hydroxide

Katy does some tests on a salt. These are her results.

Test	Observation
Add dilute sodium hydroxide	White precipitate that does not dissolve in excess
Add dilute acid	Fizzing

a Use the table to identify the salt that Katy tests. .... **[2 marks]**

b Katy adds dilute sodium hydroxide to a zinc salt. She compares her results to the results in the table.

Give **one** similarity and **one** difference Katy will see when she tests a zinc salt compared to the results in the table.

....

....

.... **[2 marks]**

c The precipitate does not dissolve in excess sodium hydroxide. What word can be used to describe a precipitate that does not dissolve? Put a ring around the correct answer.

ionic   insulator   insoluble   soluble   solution   **[1 mark]**

d Write an ionic equation, with state symbols, for the reaction that happens when dilute sodium hydroxide is added to the salt.

.... **[2 marks]**

Giant molecules and metals

1 The table shows information about the percentages of the elements in the Earth's crust.

Element	oxygen	silicon	aluminium	iron	other elements
Percentage in the Earth's crust	47	28	8	5	

a What percentage of the Earth's crust is other elements? .... **[1 mark]**

b Quartz is a common mineral. It contains the compound silicon dioxide.

Which elements does silicon dioxide contain?

.... **[1 mark]**

c Iron and aluminium are both extracted from their oxides.

What type of reaction happens when a metal is extracted from its oxide? Put a ring around the correct answer.

conduction **neutralisation** **oxidation** **reduction** **[1 mark]**

G–E

2 Copper extraction causes environmental problems due to large amounts of waste rock.

Explain why copper mining produces so much waste rock.

....

....

.... **[2 marks]**

B–A*

3 The diagrams show the structures of diamond and carbon dioxide.

Carbon dioxide

Diamond

a Carbon dioxide has a simple molecular structure and diamond has a giant covalent structure.

Describe the main differences between these two types of structure.

....

....

.... **[3 marks]**

D–C

b Tick (✓) to show which of the following statements about carbon dioxide and diamond are true and which are false.

	True	False
i Both carbon dioxide and diamond conduct electricity.	☐	☐
ii Carbon dioxide has a lower melting point and boiling point than diamond.	☐	☐
iii The atoms in both substances are held together by shared electrons.	☐	☐
iv Each carbon atom in both substances are bonded to four other atoms.	☐	☐

[4 marks]

Equations, masses and electrolysis

1 Iron is extracted from iron ore in a blast furnace. In the furnace, carbon monoxide reacts with iron oxide to make iron: iron oxide + carbon monoxide → iron + carbon dioxide.

Tick (✓) to show which substances are reactants and which are products in the reaction.

	iron oxide	carbon monoxide	iron	carbon dioxide
reactant				
product				

[1 mark]

2 Copper can be extracted from copper oxide by heating with carbon.

Complete the word and symbol equations for this reaction by filling in the missing names and formulae.

copper oxide   +   carbon   →   +

   2CuO   +   C   →   2Cu   +

[2 marks]

3 a The table shows the relative atomic masses of some elements and the relative formula masses of some compounds.

Complete the table by filling in the missing information. Use the Periodic Table to help you.

Name	Formula	Relative atomic masses of each element	Relative formula mass
sodium chloride	NaCl	Na: 23 Cl: 35.5	
magnesium chloride	$MgCl_2$	Mg: Cl: 35.5	95
calcium sulfate	$CaSO_4$	Ca: 40 S: 32 O: 16	

[3 marks]

b What is the gram formula mass of magnesium chloride?

.... [1 mark]

4 Aluminium is extracted from aluminium oxide (Al_2O_3) by electrolysis.

a Why is aluminium not extracted by reacting its oxide with carbon? Put a tick (✓) in the box next to the correct answer.

  i   The melting point of aluminium oxide is high. ☐

  ii Aluminium is a very reactive metal. ☐

  iii Aluminium oxide is a very reactive compound. ☐

  iv The density of aluminium is too low. ☐

[1 mark]

b Calculate the percentage of aluminium in aluminium oxide.

....

.... [2 marks]

c The equation for the reaction that happens at the positive electrode is: $2O^{2-} \rightarrow O_2 + 4e^-$

  i Explain why oxygen forms at the positive electrode.

....

....

.... [2 marks]

  ii Write an equation to show what happens at the negative electrode.

....

.... [2 marks]

Metals and the environment

1 The table shows some metals and their properties.

Metal	Melting point (°C)	Electrical conductivity	Mass of 1 cm³ of the metal in g	Corrosion resistance	Cost
aluminium	660	good	2.70	does not corrode	high
iron	1535	fair	7.90	corrodes quickly	medium
copper	1083	excellent	8.90	corrodes quickly	high

a Copper is used to make electrical wiring for homes.

  i Give **one** reason why copper is a good choice for making electrical wiring.

  .... **[1 mark]**

  ii Give **two** disadvantages of using copper for electrical wiring.

  ....

  .... **[2 marks]**

b Aluminium is used to make overhead electrical cables.

  Explain why aluminium is a good choice for making overhead electrical cables.

  ....

  .... **[2 marks]**

2 Read the information about mercury mining.

> **Mercury mining**
>
> Low-energy bulbs are used in the UK. They reduce carbon dioxide emissions from power stations because they use less energy than standard bulbs.
>
> The bulbs contain mercury. Large amounts of mercury are mined in China.
>
> People living near mercury mines need the jobs that the mines provide, but they complain that:
> - children and animals are sick and die because water containing toxic mercury runs into the drinking-water supply
> - extracting mercury from the ore gives off toxic gases that harm people living nearby.
>
> Some environmental groups think that all mercury mining should be stopped.

a Use ideas about cost and benefit to explain why we mine mercury even though it is toxic.

  ....

  ....

  ....

  ....

  .... **[4 marks]**

b Light bulbs contain metals other than mercury. Metals are used because they conduct electricity.

  Explain what happens when a metal conducts electricity.

  ....

  .... **[2 marks]**

C5 Extended response question

Silicon dioxide and diamond are both giant covalent structures.

Silicon dioxide

Diamond

Discuss the similarities and differences between the properties and structures of silicon and diamond.

✎ *The quality of written communication will be assessed in your answer to this question.*

B–A*

[6 marks]

Making chemicals, acids and alkalis

1 Jack uses petrol for his lawn mower. He buys the petrol in a can from a local garage. Petrol is very flammable.

a What hazard symbol should be shown on the can? Put a ring around the correct answer.

[1 mark] G–E

b What safety precautions should Jack take when he is handling the petrol?

....

.... [2 marks]

2 Pure acid compounds have different states at room temperature and pressure.

a Draw straight lines to connect each **pure acid compound** with its correct **state symbol**.

pure acid compound   **state symbol**

| (s) |

| sulfuric acid |

| (l) |

| citric acid |

| (g) |

| hydrochloric acid |

| (aq) | [3 marks]

D–C

b Which acid in the table reacts with calcium to form a salt with the formula $CaSO_4$?

.... [1 mark]

c What other product is made in the reaction?

.... [1 mark]

3 Tahira reacts copper carbonate with hydrochloric acid.

a Complete the word and balanced symbol equation for the reaction.

copper + hydrochloric ⟶   +   +
carbonate   acid

$CuCO_3$ + HCl ⟶ $CuCl_2$ +   +   [3 marks]

b Tahira puts a pH probe into the acid at the start of the reaction. She follows the pH changes as she adds the copper carbonate to the acid. She keeps adding copper carbonate until the reaction stops.

Describe and explain the pH changes that Tahira sees during the reaction.

....

....

....

.... [3 marks]

B–A*

c Tahira makes some copper nitrate by adding copper carbonate, $CuCO_3$, to nitric acid, HNO_3.

What is the formula for copper nitrate?

.... [1 mark]

Reacting amounts and titrations

1 The table shows information about some compounds.

Complete the table. Use the Periodic Table to find any relative atomic masses that you need.

Name	Formula	Relative formula mass
magnesium oxide		24 + 16 = 40
sodium oxide	Na_2O	
	Na_2CO_3	

[4 marks]

2 Zinc reacts with hydrochloric acid to make zinc sulfate.

The balanced symbol equation for the reaction is: $Zn + H_2SO_4 \longrightarrow ZnSO_4 + H_2$

a Explain why this equation is said to be 'balanced'.

....

....

[2 marks]

b What is the maximum mass of zinc carbonate that can be made from 130 g of zinc? (The relative atomic mass of zinc is 65.)

Use the Periodic Table to help you to find any relative atomic masses that you need.

....

....

....

[3 marks]

3 Rose wants to do a titration to find out how much sodium hydroxide is needed to neutralise 25 cm³ of hydrochloric acid. She has some dilute hydrochloric acid and an indicator. She also has some dilute sodium hydroxide in a burette.

a Describe how Rose should do the titration. Your answer should include details of what Rose should do to make sure that her results are as close to the true value as possible.

....

....

....

....

[4 marks]

Rose does further titrations using 25 cm³ samples of different concentrations of hydrochloric acid. She uses the same sodium hydroxide each time. Her results are shown below.

Acid	1	2	3	4
Volume of sodium hydroxide needed in cm³	50	25	20	25

b Which of the following statements about the results are true and which are false? Put a tick (✓) in one box in each row.

	True	False
a Acid 1 is the most concentrated acid.	☐	☐
b Acid 3 would react fastest with calcium carbonate.	☐	☐
c Acids 2 and 4 have the same concentration.	☐	☐
d Acid 4 is more dilute than acid 3.	☐	☐

[2 marks]

Explaining neutralisation & energy changes

1 Join the boxes to show which **acid** and **alkali** react together to make each salt.

Acid　　　　**Alkali**　　　　**Salt**

hydrochloric acid

calcium hydroxide　　　　potassium chloride

citric acid

sodium hydroxide　　　　calcium sulfate

nitric acid

potassium hydroxide　　　　sodium nitrate

sulfuric acid

[3 marks] D–C

2 Every reaction between an acid and an alkali makes a salt and one other product.

Give the name of this other product. .... **[1 mark]**

3 Neutralisation happens when ions from the acid react with ions from the alkali.

a Give the name and formula of the ion that is present in all acids.

Name: .... Formula: .... **[1 mark]**

b Give the name and formula of the ion that is present in all alkalis.

Name: .... Formula: .... **[1 mark]**

c Write an ionic equation for the neutralisation reaction between an acid and an alkali.

.... **[1 mark]**

4 Sulfuric acid reacts with sodium hydroxide to make sodium sulfate. The reaction is exothermic.

Complete the sentences by putting a ring around the correct words in **bold** in each line.

a During the exothermic reaction the temperature **increases** / **decreases**.

b Energy is **given out** / **taken in**.

c The reaction is an example of **neutralisation** / **combustion**.

[1 mark] G–E

5 This is the energy level diagram for the reaction between sulfuric acid and sodium hydroxide.

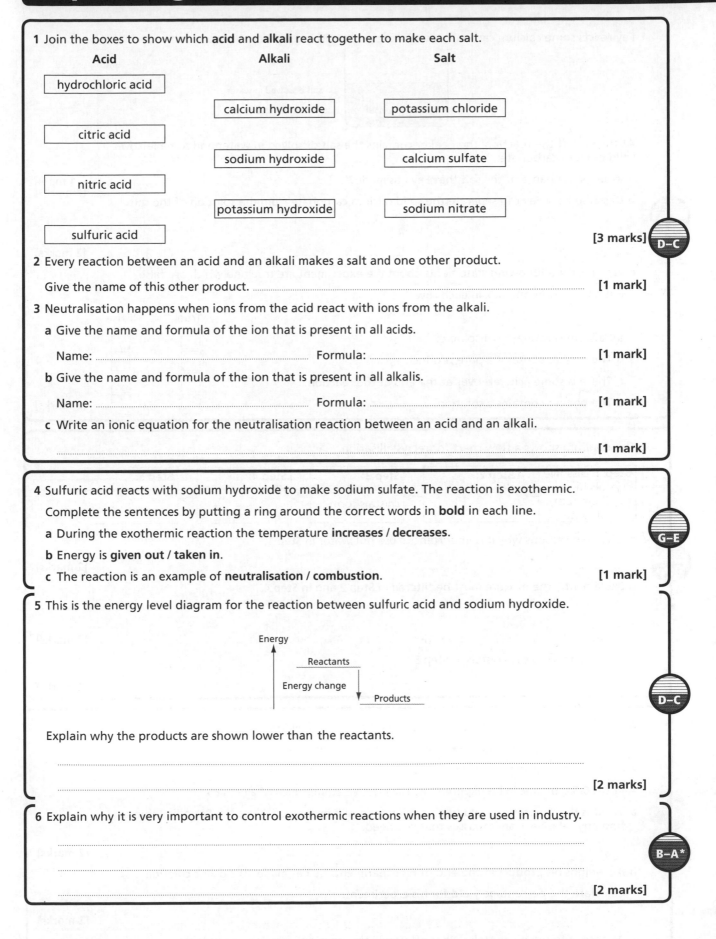

Explain why the products are shown lower than the reactants.

....

.... **[2 marks]**

D–C

6 Explain why it is very important to control exothermic reactions when they are used in industry.

....

....

.... **[2 marks]**

B–A*

Separating and purifying

1 Fay reacts some calcium carbonate with some nitric acid in a beaker to make a salt.

salt dissolved in water

solid calcium carbonate

At the end of the reaction, the beaker contains the salt dissolved in water and some leftover solid calcium carbonate.

a What is the name of the salt that Fay has made? .... **[1 mark]**

b Describe how Fay can separate the solid calcium carbonate from the solution of the salt.

....

.... **[2 marks]**

c Which of the following statements about the experiment are true and which are false?

Put a tick (✓) in one box in each row.

	True	False
a Calcium carbonate is insoluble.	☐	☐
b The beaker contains a pure product.	☐	☐
c There is some acid left over at the end of the reaction.	☐	☐
d The beaker contains a solution.	☐	☐

[2 marks]

2 The diagram shows a flow chart for recrystallisation.

Step 1: dissolve crystals in small amount of hot water	→	Step 2: filter	→	Step 3: cool until crystals form	→	Step 4: filter	→	Step 5:

a Give **two** reasons why it is important to use hot water in Step 1.

.... **[2 marks]**

b Explain why the mixture must be filtered in Step 2 and in Step 4.

....

.... **[2 marks]**

c What is done to the crystals in Step 5?

.... **[2 marks]**

3 Ali makes some zinc chloride crystals by reacting zinc with hydrochloric acid:

zinc + hydrochloric acid → zinc chloride + hydrogen

Zn + $2HCl$ → $ZnCl_2$ + H_2

Ali uses 6.5 g of zinc in his experiment. The relative atomic mass of zinc is 65.

a What is the theoretical yield of zinc chloride in Ali's experiment? Use the Periodic Table to find any relative atomic masses that you need.

.... **[2 marks]**

b Ali weighs his product at the end of his experiment. He has made 10.2 g zinc chloride.

Calculate the percentage yield for his experiment.

.... **[2 marks]**

c Ali forgot to dry his crystals. What effect will this have on his percentage yield?

Explain your reasoning.

.... **[2 marks]**

Rates of reaction

1 Ray investigates the rate of reaction between solid calcium carbonate and dilute hydrochloric acid. He investigates the effect of changing the concentration of the dilute hydrochloric acid. He carries out four experiments using different concentrations of acid. He measures the time taken to collect 20 cm³ of gas.

The table shows his results.

Experiment	Concentration of hydrochloric acid in g / dm³	Time taken to collect 20 cm³ gas in s
1	30	10
2	15	20
3	7.5	40
4	1.5	60

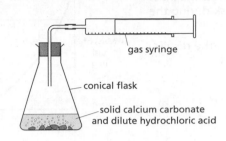

a What factors should Ray control in the experiments?

....

.... [2 marks]

b Write a conclusion to summarise what the results show.

....

.... [2 marks]

c Ray thinks that the result for Experiment 4 is an outlier. Explain why he thinks this.

....

.... [2 marks]

2 Sara does an experiment to investigate the rate of reaction between magnesium and hydrochloric acid. She follows the rate of reaction by measuring the volume of hydrogen given off. The graph shows her results.

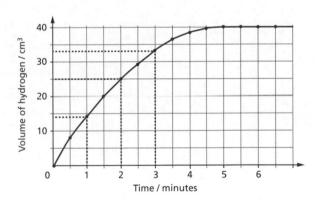

The reaction between magnesium and dilute hydrochloric acid.

a How often did Sara measure the volume of hydrogen? Circle the correct answer.

every minute   every 30 s   every second   every 6 minutes   [1 mark]

b How long did it take the reaction to finish? .... [1 mark]

c Use ideas about collisions to explain how and why the gradient of the line changes throughout the reaction.

....

....

....

[4 marks]

Joe did an experiment. He added excess solid calcium carbonate to a dilute acid in a flask.

He measured the mass of the flask during the reaction.

He plotted the results on a graph.

Mass of flask in g

Time in s

Explain how and why the shape of the graph changes during the reaction.

✎ *The quality of written communication will be assessed in your answer to this question.*

D–C

....

....

....

....

....

....

....

....

....

....

....

....

....

....

....

....

....

....

....

....

....

....

[6 marks]

The chemical industry

1 a Which of the following statements about bulk and fine chemicals are true and which are false?

Put a tick (✓) in **one** box in each row.

	True	False
Fine chemicals are not very useful.		
Fine chemicals are made on a smaller scale than bulk chemicals.		
The chemical industry only makes bulk chemicals.		
Most bulk chemicals are sold directly to the public.		

[2 marks]

G–E

b Which chemicals are made on a large scale and which are made on a small scale?

Put a tick (✓) in **one** box in each row.

	Made on a large scale	Made on a small scale
Ammonia		
Drugs		
Food additives		
Phosphoric acid		

[2 marks]

D–C

2 Which of the following roles must be done by chemists in the chemical industry?

Put a tick (✓) in the **two** correct boxes.

Driving petrol tankers	
Researching new chemical products	
Developing new processes	
Organising the payment of workers	
Designing adverts for products	

[2 marks]

3 There are strict regulations that govern the **storage** of chemicals in the chemical industry.

a Give reasons why the laws are important.

....

....

[2 marks]

B–A*

b During transport, tankers must be labelled clearly with safety information.

Suggest some information that should be included on the labels.

....

....

....

[2 marks]

Green chemistry

1 Aspirin is a drug that is used to make tablets that people take as painkillers.

a The sentences show stages in a chemical process to make aspirin. They are not in the correct order.

A Preparing the feedstocks.

B Separating aspirin from the mixture of the products.

C Checking the purity of the aspirin.

D Processing by-products and waste.

E Reacting chemicals together to make aspirin.

Put the stages in the correct order by writing the letters in the boxes.

[2 marks]

b Which stage in the process is known as synthesis?

Stage ....

[2 marks]

2 Supermarket carrier bags are usually made from polymers made from crude oil.

A new type of carrier bag is made using starch from crops of corn. The waste products produced when making the bag are biodegradable.

a Explain why the process of making the new bag is more sustainable.

....

.... [2 marks]

b What other factors need to be considered when comparing the sustainability of making the two types of bag?

....

....

.... [3 marks]

3 a Which of the following statements about the atom economy of a process are true and which are false?

Put a tick (✓) in **one** box in each row.   [2 marks]

	True	False
Green chemical processes use as low atom economy as possible.		
The atom economy of a process depends on the mass of the atoms that end up in the waste products.		
If a different reaction is used to make the same product, the atom economy is always the same.		
Reactions that use renewable feedstocks have higher atom economies.		
A reaction that makes a waste product cannot have 100% atom economy.		

b A reaction produces a waste product. A new use is found for the waste product so that it can be sold as a by-product.

Explain how this effects the sustainability of the process.   [2 marks]

....

....

Energy changes

1 Look at the diagrams for reaction A and reaction B.

Reaction A

Reaction B

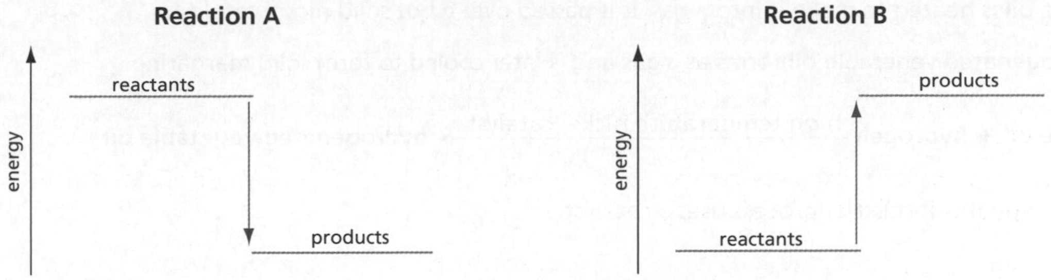

a Discuss how the diagrams show that the energy changes for the two reactions are different.

....

....

....

....

[4 marks]

b Joe measured the temperature change to the surroundings during reaction A and reaction B. He found that for one reaction the temperature of the surroundings decreased by 5 °C.

Which reaction was Joe testing? Explain how you can tell.

Reaction ....

Reason .... [1 mark]

2 The diagram shows an energy level diagram for the reaction between hydrogen and chlorine.

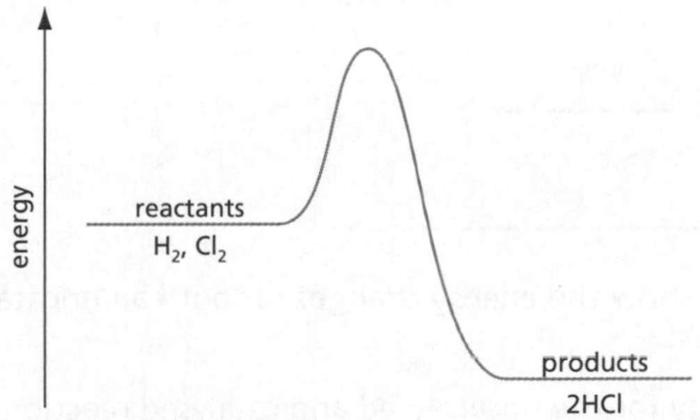

a Show the activation energy on the diagram. [1 mark]

b The reaction between hydrogen and chlorine starts when a flash of light or a spark is passed through the mixture. After it has started, it continues until the reaction has completed.

Use ideas about energy and bonds to explain why.

....

....

....

....

[4 marks]

c Hydrogen and fluorine explode when they are mixed together. A flash of light is not needed to start the reaction.

What does this tell you about the activation energy for the reaction compared to the reaction in **b**? Explain your reasoning.

....

[2 marks]

Catalysts and enzymes

1 An industrial process makes hydrogenated vegetable oils for margarines.

Vegetable oil is heated to make it into a gas. It is passed over a hot solid nickel catalyst.

The hydrogenated vegetable oil forms as a gas and is later cooled to form solid margarine.

vegetable oil + hydrogen $\xrightarrow{\text{high temperature nickel catalyst}}$ hydrogenated vegetable oil

a Explain why this industrial process uses a catalyst.

....

.... [2 marks]

b The nickel catalyst is expensive to buy but does not need to be replaced.

Explain why catalysts do not need to be replaced.

.... [1 mark]

c Nickel is a toxic metal but the hydrogenated vegetable oil is safe to eat.

Explain why the hydrogenated vegetable oil does not contain any toxic nickel.

....

.... [2 marks]

2 The diagram shows part of the energy level diagram for a reaction.

energy

reactants

products

a i Complete the diagram to show the energy changes for both an uncatalysed reaction and a reaction that uses a catalyst. [2 marks]

ii Label the activation energy for the uncatalysed and catalysed reaction. [2 marks]

b Use ideas about the energy of reacting particles to explain why catalysts make reactions happen faster.

....

....

.... [2 marks]

c The catalyst used in the reaction is an enzyme catalyst.

Describe **one** advantage and **one** disadvantage of using an enzyme catalyst in an industrial process.

....

....

.... [2 marks]

Energy calculations

1 Look at the energy level diagram for a reaction.

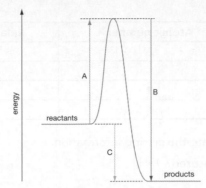

a What is the term used for the energy change shown by the arrow labelled A?

.... [1 mark] **D–C**

b Explain why this energy is needed.

....

.... [2 marks]

c Explain why this energy is different for different reactions.

....

.... [2 marks]

d Which arrow shows an energy change that happens when bonds form?

Explain your reasoning.

Energy change ....

Reason .... [1 mark]

e i What type of energy change of reaction is shown by this diagram?

.... [1 mark] **B–A***

ii Use ideas about bond energies to explain how this energy change happens.

....

.... [3 marks]

2 The equation shows a reaction that happens when carbon monoxide burns in oxygen.

$$2CO + O_2 \longrightarrow 2CO_2$$

The table shows the bond energies for the reactants and products.

Bond	Energy (kJ/mol)
$C \equiv O$ (in CO)	1 079
$O = O$	498
$C = O$ (in CO_2)	805

a Calculate the energy given out during the reaction. [3 marks] **B–A***

b How do the values of the energy changes show that this is an exothermic reaction?

....

.... [2 marks]

Reacting masses

1 The table shows some information about the symbol, atomic number and relative atomic mass of some elements.

Symbol	Atomic number	Relative atomic mass
He	2	4
		12
	9	19
Mg		24

[3 marks]

a Use the Periodic Table to complete the missing information.

b How is relative atomic mass measured?

....

.... [2 marks]

c Eve looks at the data in the completed table in **a** and has this idea.

> The data in the table shows that the relative atomic mass of an atom is always double the atomic number.

What evidence in the table supports and does not support Eve's idea?

....

....

.... [3 marks]

d What further evidence would Eve need to collect to find out if her idea works for other elements?

....

.... [2 marks]

2 Magnesium reacts with dilute hydrochloric acid.

$Mg + 2HCl \longrightarrow MgCl_2 + H_2$

a The table shows some information about the relative masses of the reactants and products in the equation. Complete the missing information. Use the Periodic Table to help you. [3 marks]

Formula	Relative mass
Mg	24
HCl	
$MgCl_2$	
H_2	

b What mass of magnesium chloride can be made from 6 g of magnesium?

Use the equation and the table of relative masses in **a** to help you.

....

....

....

.... [5 marks]

Alkanes

1 Ethane is a hydrocarbon that contains two carbon atoms.

a i What does the term 'hydrocarbon' mean?

....

.... **[2 marks]**

ii What is the formula for ethane?

Put a tick (✓) in the box next to the correct answer.

CH_4	
C_2H_6	
C_3H_7OH	
C_2H_3COOH	

[1 mark]

b Complete the word equation to show what happens when ethane burns.

ethane + ... ⟶ .... + ... **[3 marks]**

c Other than burning, ethane is unreactive.

Use ideas about bonds and bond energies to explain why ethane is unreactive.

....

....

.... **[3 marks]**

2 Propene is another hydrocarbon. It is used to make poly(propene) for making carpets.

This is the structure of propene.

$$\begin{array}{ccccc} & & H & H & \\ & & | & | & \\ H{\diagdown} & & & & \\ & C=C & - & C - H \\ H{\diagup} & & | & | & \\ & & & H & \\ & & & H & \end{array}$$

a The molecular formula for ethane is C_2H_6.

i What is the molecular formula for propene?

.... **[1 mark]**

ii Give **one** similarity and **one** difference between a molecule of ethane and a molecule of propene.

....

.... **[2 marks]**

b Ethane is saturated. Propene is unsaturated.

Explain what these statements mean.

....

.... **[2 marks]**

Alcohols

1 a Methanol and ethanol have similar uses.

   i Give **two** examples of uses that are shared by methanol and ethanol.

   ....

   .... [2 marks]

   ii Ethanol is used to make alcoholic drinks.

   Explain why methanol **cannot** be used in this way.

   .... [1 mark]

   iii Complete the equation for the combustion of ethanol by filling in the gaps.

   ethanol   + ...   ⟶   carbon dioxide   + ...

   C_2H_5OH + $3O_2$ ⟶   2 ....   3 ....   [3 marks]

2 Ethane and ethanol both contain two carbon atoms.

   Ethane and ethanol both burn in air to form carbon dioxide and water.

   a What feature of the structures of ethane and ethanol cause both molecules to be flammable?

   ....

   .... [1 mark]

   b Ben says that both molecules are hydrocarbons.

   Do you agree?

   Explain your reasoning.

   ....

   .... [2 marks]

   c Ethanol is more soluble in water than ethane.

   i Describe how **one other** physical property of ethanol is different to ethane.

   .... [1 mark]

   ii Explain why the properties of ethanol and ethane are different.

   .... [1 mark]

3 A piece of sodium was added to water, ethanol and hexane.

   Hexane is a liquid alkane.

   a Predict the similarities and differences between what you would see when sodium was added to each compound.

   ....

   ....

   ....

   ....

   .... [4 marks]

   b Name the element and compound that forms in the reaction between sodium and ethanol.

   Element...

   Compound.. [2 marks]

Fermentation and distillation

1 Fermentation is used to make beer.

Use these words to answer the questions.

sugar   carbon dioxide   yeast   ethanol   water

a What microorganism is needed for respiration?

....   [1 mark]

b What is used for food by the microorganism?

....   [1 mark]

c When making beer, what is the useful product of fermentation?

....   [1 mark]   **G–E**

d What is the waste product?

....   [1 mark]

e The products of fermentation can be distilled to make whisky.

Explain how distillation works.

....

....

....   [3 marks]

2 Alex makes bread. Making bread uses the process of fermentation to make a gas that causes the bread to rise.

The box shows part of a recipe for making bread.

- Add sugar to bread mixture.
- Add yeast to bread mixture.
- Cover mixture with plastic film.
- Put in warm place to rise.

a Explain why each stage of the recipe is needed to make sure that the bread rises.

....

....

....

....   [4 marks]

D–C

b What gas causes the bread to rise?

....   [1 mark]

3 Joe works as a wine seller.

He knows that wine made by fermentation contains up to about 13.5% ethanol.

Explain why fermentation cannot be used to make alcoholic drinks with much higher concentrations of ethanol.

B–A*

....

....   [3 marks]

Alternatives to fermentation

1 Which raw materials can be used to make ethanol?

Put a tick (✓) in the boxes next to the **two** correct answers.

G–E

Waste straw from crops	
Nitrogen and hydrogen in the air	
Calcium carbonate in limestone	
Crude oil	
Sulfur extracted from volcanoes	

[2 marks]

2 Ethanol is made from ethane in two stages.

Stage 1 : $C_2H_6 \longrightarrow C_2H_4 + H_2$

Stage 2 : $C_2H_4 + H_2O \longrightarrow C_2H_5OH$

C–D

a Write a word equation for each stage of the process.

....

.... [2 marks]

b Explain why making ethanol in this process uses large amounts of energy.

.... [1 mark]

c The process has a high atom economy.

i Explain how the equations show that this is true.

....

....

....

.... [3 marks]

ii Suggest **one** use for the by-product of the process.

.... [1 mark]

B–A*

3 Ethanol can be made from plant material either by fermentation or by using *E.coli* bacteria.

a Discuss the similarities and differences between these two processes.

....

....

....

....

.... [4 marks]

b Give a disadvantage of using each process to make ethanol.

....

.... [2 marks]

Carboxylic acids

1 a Complete the table of information about some carboxylic acids.

Name	Formula	Use
	HCOOH	Preservative in animal feed
Ethanoic acid		

[3 marks]

b Carboxylic acids can be used as preservatives for food.

Some properties of carboxylic acid mean that using them in food makes the food less attractive to eat.

Give **two** properties of carboxylic acids that may make food less attractive to eat.

....

....

[2 marks]

2 The diagram shows the structure of ethanoic acid.

a Put a (ring) around the carboxylic acid functional group in the molecule. [1 mark]

b Ethanoic acid reacts with magnesium.

Complete the word equation for the reaction

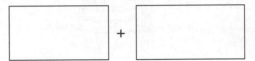

ethanoic acid	+	magnesium	→		+	

[2 marks]

c Ethanoic acid reacts with another chemical to give sodium ethanoate and water.

What is the name of the chemical?

....

[1 mark]

d Complete the table of information about some carboxylic acid salts.

Name of salt	Formulae	Ions present
	HCOONa	HCOO⁻ and Na⁺
Sodium ethanoate		
Magnesium ethanoate	$(CH_3COO)_2Mg$	

[4 marks]

3 These are ionic equations for what happens when ethanoic acid and hydrochloric acid react as acids.

$CH_3COOH(aq) \rightarrow CH_3COO^-(aq) + H^+(aq)$

$HCl(aq) \rightarrow H^+(aq) + Cl^-(aq)$

a Explain how these equations show that both ethanoic acid and hydrochloric acid are behaving as typical acids.

....

[1 mark]

b Explain what the symbol (aq) means in each reaction.

....

[1 mark]

Carboxylic acids

1 Carboxylic acids are used to preserve animal feed.

  a Give **two** examples of carboxylic acid.

    ....

    .... **[2 marks]**

  b Carboxylic acids are safe to use in animal feed at low concentrations.

    Explain why it is important that the concentration of carboxylic acid is kept low.

    ....

    .... **[2 marks]**

  c Sulfuric acid cannot be added to animal feed.

    Explain why carboxylic acids are safe to use in food, but sulfuric acid is not.

    ....

    ....

    ....

    .... **[3 marks]**

2 Liz has bottles of two acids.

Ethanoic Acid   Hydrochloric Acid

Liz thinks that the acids have different pH values.

  a Describe how Liz could test the pH and what results she should expect.

    ....

    .... **[2 marks]**

  b Liz adds sodium carbonate to each acid.

    What are the similarities and differences between the way the two acids react with sodium carbonate?

    ....

    ....

    .... **[3 marks]**

3 Nitric acid (HNO_3) and methanoic acid (HCOOH) both dissolve in water to produce hydrogen ions.

The equation shows what happens when nitric acid produces hydrogen ions.

$HNO_3(aq) \longrightarrow H^+(aq) + NO_3^-(aq)$

  a Write a similar equation to show what happens when methanoic acid dissolves in water.

    .... **[2 marks]**

  b Use the equations to explain why nitric acid is a strong acid and methanoic acid is a weak acid.

    ....

    ....

    .... **[3 marks]**

Esters

1 Eve is talking about esters.

> I have been learning about esters. Flowers are the only natural source of esters. Esters give flowers their strong colours.

a Some of Eve's ideas about esters are not correct.

Rewrite the statements so that they are correct.

....

....

.... **[3 marks]**

b Esters can be made synthetically.

Synthetic esters are used in nail varnish remover, paints and inks.

i Why are esters useful to make these products?

.... **[1 mark]**

ii Products containing esters have a flammable symbol on their label.

Give two reasons why you would expect esters to be highly flammable.

....

.... **[2 marks]**

c Esters are also added to some plastics that are used as packaging.

Explain why.

....

.... **[2 marks]**

d Some esters are naturally occurring.

Suggest a reason why more products are made from synthetic esters than from natural esters.

.... **[1 mark]**

2 Ethyl ethanoate is made from a carboxylic acid and an alcohol.

a Complete the equation for the reaction by filling in the boxes.

$$\boxed{} + \boxed{} \longrightarrow \text{ethyl ethanoate} + \boxed{}$$

$$CH_3COOH + \boxed{} \longrightarrow CH_3COOC_2H_5 + \boxed{}$$

[4 marks]

b The reaction is too slow at room temperature.

Give the conditions that are used to make the reaction faster.

....

.... **[3 marks]**

Making esters

1 Joe makes some ethyl ethanoate.

  a Joe uses some of the following stages. The stages are not in the correct order. There is one extra stage.

  A Drying
  B Distillation
  C Reflux
  D Titration
  E Reactions in a tap funnel

  i Put the stages that Joe used in the correct order.

  Write the letters in the boxes. [2 marks]

  ii Concentrated sulfuric acid and one other acid are added during the process.

  Which stage happens immediately after the acids are added?

  .... [1 mark]

  iii Complete the table to show the information about the two acids.

Name of acid	Formula	Purpose
concentrated sulfuric acid		
	CH_3COOH	Needed as a reagent for the reaction

  [3 marks]

  b Using a tap funnel in stage **E** involves separating two liquids from each other and separating liquids from a gas.

  Explain how a tap funnel can be used to do each of these separations.

  ....

  ....

  .... [2 marks]

  c Give the name of the substance that is added in stage **A**.

  .... [2 marks]

2 Reflux and distillation are both techniques that are involved when liquid esters are made.

  What are the similarities and differences between reflux and distillation?

  Your answer should include information about how each technique is carried out and the reasons for using each technique.

  ....

  ....

  ....

  ....

  ....

  .... [5 marks]

B–A*

Fats and oils

1 Which of the following statements about fats and oils are true and which are false?

Put a tick (✓) in **one** box in each row.

	True	False
Fats are solids at room temperature.		
The melting point of oils is above room temperature.		
Oils for use in foods are usually made from animals.		
Living things use fats and oils for energy.		

[2 marks]

2 The diagram shows the structure of an oil molecule, oil A.

Oil A

a Put a ring around an ester group in the structure. [1 mark]

b Why are fats and oils called **triesters**?

.... [1 mark]

c The molecule of oil A can be hydrolysed.

This means reacting the ester groups with water to make an alcohol molecule and three carboxylic acid molecules.

i Which small molecule is added during this reaction?

Put a ring around the correct answer. [1 mark]

CO_2 C_2H_5OH CH_3COOH HCl H_2O

ii Draw the **structure** and give the **name** of the alcohol molecule that is formed by hydrolysing oil A.

Name .... [3 marks]

d Olive oil is another type of oil.

Which statement describes the difference between olive oil and oil A?

Put a tick (✓) in the box next to the correct answer.

They are made from different alcohols.	
Olive oil does not contain ester groups.	
They are made from different fatty acids.	
Only oil A is a hydrocarbon, olive oil is not.	

[1 mark]

e Oil A is a saturated oil. Olive oil is an unsaturated oil. Explain what this means.

....

.... [2 marks]

Reversible reactions

1 Fizzy drinks contain carbon dioxide gas dissolved in water.

The equation shows what happens when carbon dioxide dissolves in water.

$CO_2(g) \rightleftharpoons CO_2(aq)$

a What does the symbol $\rightleftharpoons$ mean? ....

.... **[1 mark]**

b After the bottle of fizzy drink has been opened, the drink goes 'flat'.

This is because the carbon dioxide dissolved in the water leaves the drink and turns back into carbon dioxide gas.

Explain how the equation shows that this can happen.

....

.... **[2 marks]**

c Fizzy drinks do not go 'flat' if the top of the bottle is kept tightly closed.

Use ideas about equilibrium to explain why.

....

.... **[2 marks]**

2 Hydrogen reacts with chlorine to make hydrogen chloride.

The reaction is reversible.

$H_2(g) + Cl_2(g) \rightleftharpoons 2HCl(g)$

The equilibrium can be studied if the reaction is done in a closed container.

a Explain why it is important that the reaction is done in a closed container.

....

.... **[2 marks]**

b After some time the reaction reaches dynamic equilibrium.

Use ideas about rate of reaction to explain what this means.

....

....

.... **[3 marks]**

c The concentrations of hydrogen, chlorine and hydrogen chloride were measured during the reaction and plotted on a graph.

i What happens at A?

.... **[1 mark]**

ii Explain how and why the concentrations of hydrogen and chlorine change during the reaction.

....

....

.... **[3 marks]**

The Haber process

1 Until about 100 years ago only natural processes could fix nitrogen.

Farmers relied on animal waste to provide fertiliser for their crops.

Fritz Haber invented a new process for making synthetic nitrogen compounds.

a What does 'fixing' nitrogen mean?

Put a tick (✓) in the box next to the correct answer.

Dissolving nitrogen gas in water.	
Breaking nitrogen compounds down to make plant protein.	
Reacting the element nitrogen to make a nitrogen compound.	
Putting nitrogen gas under pressure so that it turns to a liquid.	

[1 mark]

b Why do farmers need to add nitrogen-based fertilisers to their soils?

....

.... **[2 marks]**

2 The graph shows the effect of changing the conditions on the yield of ammonia in the Haber process.

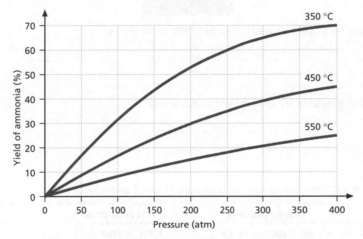

a How does the yield of ammonia change when the temperature is increased?

.... **[1 mark]**

b How does the yield of ammonia change when the pressure is increased?

.... **[1 mark]**

c i What conditions give the highest yield of ammonia?

Put a ring around the **two** correct answers.

low temperature   **high temperature**   **low pressure**   **high pressure** **[1 mark]**

ii Explain why the conditions that give the highest yield are **not** used when the process is done on an industrial scale.

....

.... **[2 marks]**

d In industry, the process uses a 'finely divided iron catalyst'.

Explain why the catalyst is 'finely divided'.

....

.... **[2 marks]**

Alternatives to Haber

G–E

1 a Which of the following are true only for **synthetic fertilisers**, true only for **organic fertilisers** or true for both?

Put a tick (✓) in **one** box in each row.

	True only for synthetic fertilisers	True only for organic fertilisers	True for both
Contain nitrogen compounds			
Made from feedstocks			
Made from waste material			

[3 marks]

b Ben buys some organic vegetables. The vegetables are grown on a farm that uses fertilisers from animal waste rather than synthetic fertilisers.

I like to buy these vegetables because synthetic fertilisers cause problems in local rivers. Organic fertilisers do not.

i What problems do fertilisers cause if they get into local rivers?

....

.... [2 marks]

ii Do you agree with Ben's view? Explain your reasoning.

....

.... [2 marks]

D–C

2 The table shows some information about two processes used to make fertiliser.

Process	Haber process	Enzyme process
Raw materials	Natural gas, hydrogen, air	Air, water
Conditions	High temperature, high pressure, use of catalyst	Room temperature, use of enzyme from genetically modified bacteria
Scale of use	Industrial scale	Experimental laboratory scale

a The Haber process uses more non-renewable resources than the enzyme process. Explain why.

....

....

....

....

.... [5 marks]

b What is the main disadvantage of the enzyme process?

.... [1 mark]

B–A*

3 Joe says that he does not think that using animal waste as fertiliser for growing vegetables is a sustainable process in the UK on a large scale. Explain why he is right.

....

.... [2 marks]

Analysis

1 Jack works for a company that makes metal compounds. The company has a factory near a river.

Jack checks the water quality in the river to make sure that the factory is not affecting the water.

a The table shows some of the data that Jack collects.

Which data is quantitative data and which is qualitative data?

Put a tick (✓) in the correct box in each row.

	Quantitative data	Qualitative data
What metal ions the water contains		
Concentration of dissolved oxygen		
Water temperature		
Types of algae in the water		
Rate of flow of the river		

[2 marks]

b The factory operates five days a week.

Jack wants to find out if the factory affects water quality while it is operating.

Jack collects samples from the river to take back to his laboratory for analysis.

i How should Jack choose the samples?

....

....

....

[2 marks]

ii What should Jack do to make sure that the samples do not get contaminated?

....

....

....

[2 marks]

c Jack tests his samples in the laboratory.

During testing he makes sure that he:

• Repeats each test several times for each sample.
• Follows a standard procedure each time.

Explain why he does this.

....

....

....

[2 marks]

d Jack wants to keep some of the samples he collects for testing a few days later.

He wants to collect quantitative data about the amount of single-celled algae in the water.

The box shows some information about this type of algae.

> • Algae are single-celled plants.
> • They grow and reproduce in light, warm conditions.

How should Jack store the samples of water to make sure that the results of his tests will be accurate?

....

[2 marks]

Principles of chromatography

1 Join the boxes to link each **term** with its correct **meaning**.

Term	Meaning
Aqueous solution	A substance that dissolves in a liquid.
Non-aqueous solution	Substances dissolved in water.
Solvent	Substances dissolved in liquids other than water.
Solute	A liquid used to dissolve substances.

[2 marks]

2 Ella does an experiment. She shakes some solid iodine with water. She shakes some solid iodine with hexane. She shakes some solid iodine with a mixture of water and hexane and leaves the layers to settle.

The table shows her results.

	Results
Iodine with water	Pale brown solution with solid left at the bottom.
Iodine with hexane	Deep purple solution.
Iodine with water and hexane	Deep purple upper layer, pale brown lower layer.

a What conclusions can you make from Ella's results about the distribution of iodine between hexane and water? Show how you used the results to make your conclusions.

....

....

....

....

[4 marks]

b What conclusions can you make from Ella's results about the density of hexane compared to water?

Explain your reasoning.

....

....

[2 marks]

3 Fay drops water onto a spot of ink on a filter paper. The ink spreads out to make rings.

She repeats the experiment with another spot of the same type of ink. This time she drops alcohol onto the spot.

The two solvents give different results.

Explain why the results are different.

Use ideas about the relative solubility of the substances in the inks in the mobile phase and the stationary phase in your answer.

Results using water — filter paper — Results using ethanol

....

....

....

....

[4 marks]

Paper and thin-layer chromatography

1 The statements describe some steps needed to set up and run a paper chromatogram.

There is one incorrect step.

The steps are not in the correct order.

A	Wait for the solvent to almost reach the top of the paper.
B	Put a small spot on a piece of chromatography paper.
C	Heat the solvent.
D	Put the paper in the solvent.
E	Dry the paper.

Put the steps in the correct order by writing the letters in the boxes.

[2 marks]

2 Guy works for a laboratory that tests food colourings.

Some dyes are banned from being used in foods.

Guy runs a chromatogram of some food colourings (**X** and **Y**), some safe dyes (**A**, **B** and **C**) and a banned dye (**D**).

The diagram shows his results.

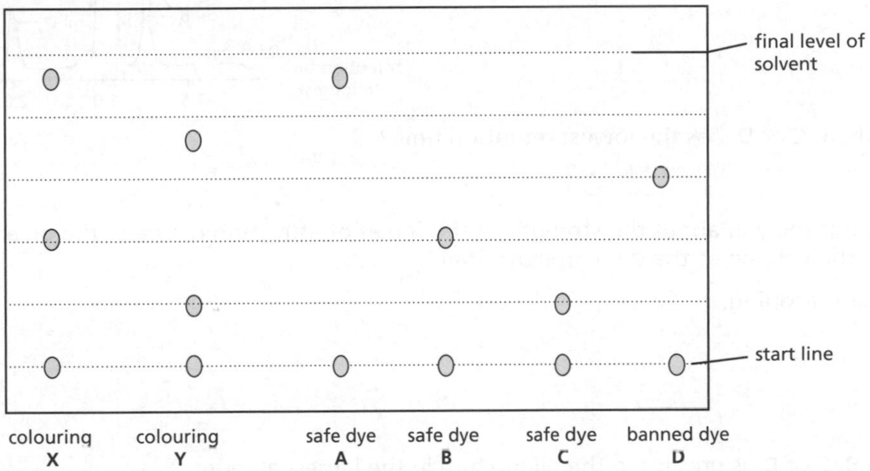

final level of solvent

start line

| colouring X | colouring Y | safe dye A | safe dye B | safe dye C | banned dye D |

a What conclusions can Guy make about the safety of food colourings X and Y? Explain your reasoning.

....

....

....

....

....

[4 marks]

b Calculate the R_f value of the banned dye.

Show your working.

....

....

....

[3 marks]

c Some chromatograms involve using locating agents.

Why are locating agents sometimes necessary?

....

[1 mark]

Gas chromatography

1 a Complete the table to show the **parts** of a gas chromatogram and its **purpose**.

Part	Purpose
Carrier gas	
	Acts as stationary phase.
Oven	
Detector	
	Gives a graph showing a peak for each substance.

[5 marks]

b Nitrogen is often used as a carrier gas in a gas chromatogram.

Explain why this is a good choice.

....

....

[2 marks]

2 The printout shows the results of a gas chromatogram of a sample of a margarine that is made by a company.

The margarine contains several different oils.

a i Which oil, **A**, **B**, **C** or **D**, has the lowest retention time?

....
[1 mark]

ii What does this tell you about the strength of the forces of attraction between the oil and the solid in the column of the gas chromatogram?

Explain your reasoning.

....

....
[2 marks]

b Which oil, **A**, **B**, **C** or **D**, is present in the oil mixture in the largest amounts?

Give a reason for your answer.

Oil ....

Reason ....
[2 marks]

c Another company makes a similar margarine from plant oils.

A researcher analyses a sample of margarine. He wants to know which company made the margarine.

How can the researcher use gas chromatography to find out which company made the margarine?

....

....

....

....
[4 marks]

d The researcher uses a chromatogram that has a mass spectrometer attached to the detector. Explain why this is useful.

....

....
[2 marks]

Quantitative analysis

1 Jay makes up a standard solution. He writes down his method.

His method contains errors.

    Step 1 Weigh out 20 g of solid.

    Step 2 Put the solid in a beaker.

    Step 3 Use a measuring cylinder to measure out 100 cm³ water.

    Step 4 Add the water to the beaker and stir.

a A measuring cylinder is not accurate enough to make a standard solution.

What piece of apparatus should Jay use to measure volume when he is making his solution?

.... [1 mark]

b Rewrite Jay's method to show how to accurately make a standard solution.

....

....

....

.... [4 marks]

2 Katy wants to find out the amount of dissolved solids in seawater near where she lives.

She collects six separate samples. She leaves the water to evaporate and weighs the solid left behind.

The table shows her results.

Sample	1	2	3	4	5	6
Mass of solid in g	3.5	3.4	3.6	4.6	3.3	3.7

a i Which reading is an outlier?

.... [1 mark]

ii Calculate the best estimate of the amount of solids in the seawater.

[2 marks]

b Katy collects another set of samples. These are the results for the second set of samples.

Sample	7	8	9	10	11	12
Mass of solid in g	3.4	3.5	3.6	3.7	3.5	3.3

i What is the range of the results for the second set of samples?

.... [1 mark]

ii What evidence is there to suggest that both sets of Katy's results are accurate?

....

.... [2 marks]

3 The table shows information about some solutions.

Complete the table by filling in the missing information.

Solution	Mass of solute in g	Volume of solution	Concentration in g/dm³
A	20	0.5 dm³	
B	10	200 cm³	
C		250 cm³	12

[3 marks]

Acid-base titration

1 Nia is a laboratory technician. She has a bottle of dilute sodium hydroxide and she wants to find out its concentration. She does a titration to find out.

Write a set of instructions for Nia to follow to do her titration. Your answer should include what she needs to use and how she should do the titration.

....

....

....

....

[5 marks]

2 Olly does a titration. He measures the amount of acid needed to neutralise some dilute sodium hydroxide. The table shows his results.

Sample	Rough	Accurate 1	Accurate 2	Accurate 3	Accurate 4	Accurate 5
Burette reading in cm³	23.0	22.5	22.6	22.4	22.4	22.6

Suggest reasons why Olly's results vary.

....

....

[2 marks]

3 Pat does a titration to find out how much sodium carbonate is in a sample of baking soda. She puts the sample in a conical flask and does a titration using an indicator and hydrochloric acid. The concentration of the hydrochloric acid is 11 g/dm³ and she uses 25.0 cm³ to neutralise the sodium carbonate.

This is the equation for the reaction.

$Na_2CO_3 + 2HCl \longrightarrow NaCl + H_2O + CO_2$

a Pat notices that the contents of the flask fizz during the reaction.

Explain why this happens.

....

....

[2 marks]

b Calculate the relative formula mass of sodium carbonate and hydrochloric acid.

Use the Periodic Table to help you.

....

....

[2 marks]

c Pat calculates the mass of hydrochloric acid used in the titration.

Some of her calculation is shown.

Complete the calculation by filling in the boxes.

Mass of HCl used = ☐ × $\dfrac{25.0}{\boxed{}}$

Answer = ☐ g

[2 marks]

d Use the equation and your answers to b and c to work out the mass of sodium carbonate in the original sample.

....

....

....

[3 marks]

C7 Extended response question

Rana tests the reactions of some acids.

Her results are shown in the table.

Acid	Reaction with sodium carbonate	Reaction with magnesium	pH
ethanoic acid	slowly fizzes	slowly fizzes, magnesium disappears very slowly	3.5
hydrochloric acid	very fast reaction, violent fizzing	very fast reaction, magnesium disappears very quickly	1
sulfuric acid	very fast reaction, violent fizzing	very fast reaction, magnesium disappears slowly	1
citric acid	fizzes	fizzes, magnesium disappears slowly	3

What conclusions can you make about the strength of the acids in the table?

Explain your reasoning.

🖋 *The quality of written communication will be assessed in your answer to this question.*

[6 marks]

C1 Grade booster checklist

I know that the Earth's atmosphere was probably formed by gases from volcanoes.	
I know that power stations and vehicle use both add to air pollution.	
I can explain that hydrocarbons only contain carbon and hydrogen.	
I know that incomplete combustion produces carbon monoxide and carbon particulates.	
I understand that carbon monoxide, sulfur dioxide and nitrogen oxides are pollutants.	
I know that outliers are results that are different from all the others.	
I am working at grades G/F/E	

I understand that gas particles are very small with lots of empty space between them.	
I can recall that the Earth's atmosphere has changed over time.	
I know that solid particulates (soot) are released by both natural and man-made processes.	
I understand that sulfur dioxide and nitrogen oxides are pollutants that make acid rain.	
I know that hydrocarbon fuels burn in oxygen to form water and carbon dioxide.	
I know that oxidation is adding oxygen and reduction is removing oxygen.	
I understand that atoms are rearranged during chemical reactions to make new products.	
I know that nitrogen oxides are formed in hot car engines from nitrogen and oxygen in the air.	
I can explain that range is the difference between high and low results, and mean is the average.	
I can work out the true value from a set of results.	
I am working at grades D/C	

I know that explanations are evidence-based, but can change if new evidence is found.	
I understand that a correlation is a link between a factor and an outcome.	
I know that a causal link needs evidence showing that one factor always causes an outcome.	
I know that mass is always conserved in a chemical reaction.	
I can explain that NO is oxidised to NO_2, and jointly these are referred to as Nx.	
I know the benefits and problems of using biofuels and electric cars.	
I am working at grades B/A/A*	

C2 Grade booster checklist

I know that rubber, plastic and metals are useful materials that have different properties.	
I know that hydrocarbons only contain carbon and hydrogen atoms.	
I understand that crude oil is a mixture of thousands of different hydrocarbons.	
I know that crude oil can be separated by fractional distillation into fractions of similar-sized molecules.	
I know that monomers join up to make polymers.	
I understand that synthetic materials such as plastic have replaced some natural materials.	
I know that very small silver nanoparticles kill bacteria.	
I am working at grades G/F/E	

I understand that the properties of different materials need to be considered when choosing one for a job.	
I know that cotton, paper, silk, wool, iron ore and limestone are all natural materials.	
I know that synthetic materials are manufactured using raw materials from the Earth's crust.	
I understand that a fraction of crude oil contains similar-sized molecules with similar boiling points.	
I know that chain length, cross-linking and plasticisers alter the strength and flexibility of plastics.	
I know that nanotechnology makes molecules up to 100 nm in size, and that these have different properties to the same large-scale material.	
I am working at grades D/C	

I know that synthetic materials can be designed to provide the properties needed for a particular purpose.	
I know that the boiling point of a hydrocarbon molecule is linked to the intermolecular forces between molecules, and these forces increase with chain length.	
I understand that increasing crystallisation can be done by reducing the number of branches, giving a more regular pattern of aligned molecules, making a polymer stronger and increasing melting point.	
I know that carbon spheres containing 60 atoms are called 'buckyballs', which can be made into very strong nanotubes, and that this is a developing technology with possible unknown risks.	
I know that nanoparticles are very strong due to their high surface area to volume ratio.	
I am working at grades B/A/A*	

C3 Grade booster checklist

I understand that geological changes happened in Britain by the slow movements of tectonic plates.	
I know that coal, salt and limestone are important raw materials.	
I know that salt is used in the food industry as a flavouring and as a preservative.	
I understand that alkalis neutralise acid soils, are used for dying clothes, and for making glass and soap.	
I know that chlorine is used to kill microbes in drinking water.	
I know that electrolysis breaks up sodium chloride solution into chlorine, hydrogen and sodium hydroxide.	
I can explain that plasticisers are small molecules added to PVC to make it more flexible.	
I am working at grades G/F/E	

I know that Britain has experienced different climates and has rocks from different ancient continents.	
I understand how limestone, salt and coal formed.	
I know that solution mining extracts sodium chloride solution for use in industry, but that too much is bad.	
I know that a diet containing excess salt is linked to high blood pressure and heart failure.	
I can recall that the first alkali production created harmful by-products.	
I know that there are health concerns associated with plasticisers.	
I understand that a Life Cycle Assessment (LCA) measures the energy use and environmental impact over the life of a product, from cradle to grave.	
I am working at grades D/C	

I know that magnetic clues in igneous rocks can be used to track continental movements.	
I understand that fossils, shell fragments, grain shape and ripple marks give evidence about the conditions when sedimentary rocks formed.	
I know that salt extraction can cause subsidence and environmental problems.	
I can explain what the DH and Defra are, and what they do.	
I know that soluble hydroxides and carbonates are alkalis, and can predict the products when they react with acids.	
I understand that some people disapprove of chlorination, but the benefits outweigh the risks.	
I know the difference between perceived and calculated risk.	
I am working at grades B/A/A*	

C4 Grade booster checklist

I understand that elements are arranged into patterns in the Periodic Table.	
I know that an atom consists of a nucleus containing protons and neutrons, with electrons arranged in shells around the outside.	
I know that the first electron shell holds 2 electrons and the second shell holds 8 electrons.	
I know that a horizontal row across the Periodic Table is called a Period.	
I know that a vertical column in the Periodic Table is called a Group.	
I know the colours and states of the halogens (Group 7) at room temperature and as gases.	
I know that the halogens react with Group 1 metals and iron.	
I understand that ionic compounds contain charged particles and conduct electricity when they are molten or dissolved in water.	
I am working at grades G/F/E	

I know that Döbereiner, Newlands and Mendeleev were three scientists who had different ideas about how to arrange elements into patterns.	
I know that each element has a unique flame colour and line spectrum.	
I understand that the proton number of an atom gives the number of electrons for the atom.	
I know that for the first 20 elements, the third electron shell in an atom holds 8 electrons.	
I know that proton numbers, numbers of electrons and properties change across a period.	
I know that Group 1 metals have trends in their physical properties.	
I can explain how Group 1 metals react with water and chlorine.	
I know that the halogens have trends in their physical properties.	
I understand that halogens contain diatomic molecules (molecules that contain two atoms).	
I know that the halogens get less reactive down the group.	
I know that more reactive halogens can displace less reactive halogens from their compounds.	
I understand that ionic compounds conduct electricity when their ions are free to move.	
I am working at grades D/C	

I can use the Periodic Table to work out the number of protons, neutrons and electrons in an atom.	
I know that the electron arrangement of the atoms in an element is linked to its position in the Periodic Table.	
I know that the electron arrangement in atoms is linked to the reactivity of the element.	
I know that Group 1 elements are more reactive when they have more electron shells and Group 7 are less reactive when they have more electron shells.	
I understand that positive ions form when atoms lose electrons and negative ions form when they gain electrons, and that the formula of an ionic compound contains positive and negative ions with a balance of charges.	
I am working at grades B/A/A*	

C5 Grade booster checklist

I know that air contains elements and compounds that contain non-metal atoms, and I know the main gases in the air.	
I know that the Earth's hydrosphere contains water with dissolved ionic salts.	
I know that ionic salts can be identified by testing with dilute sodium hydroxide, dilute acid, dilute silver nitrate and dilute barium chloride.	
I understand that the lithosphere contains rocks, minerals and ores.	
I know that metals can be extracted from metal oxides by reduction (taking away oxygen).	
I understand that electrolysis breaks down electrolytes when an electric current passes through.	
I know that metals are strong, malleable, have high melting points and conduct electricity.	
I am working at grades G/F/E	

I know that molecules in the air can be shown in either 2D or 3D.	
I know that molecules in the air have low melting points and boiling points because there are weak attractions between the molecules.	
I know that in ionic compounds there are strong forces between positive and negative ions called ionic bonds, and that these determine the properties of the compound, such as melting point and electrical conductivity.	
I understand that tests for positive and negative ions depend on the formation of insoluble precipitates.	
I know that large amounts of rock have to be mined to produce small amounts of metals.	
I know that diamond, graphite and silicon dioxide are giant covalent structures with similar properties.	
I know that relative atomic mass can be used to work out relative formula mass and relative gram mass.	
I understand that, during electrolysis, metals form at the negative electrode and non-metals form at the positive electrode.	
I know that metal processing harms the environment because of toxic waste and large amounts of waste rock.	
I am working at grades D/C	

I know that molecules in the air are held together by sharing electrons in covalent bonds.	
I know that the positive and negative charges on ions balance in the formula of an ionic compound.	
I know that precipitation reactions can be represented using ionic equations.	
I know that redox reactions involve both reduction and oxidation.	
I know that the bonding and structure of diamond and graphite are linked to their properties.	
I know that relative gram mass can be used to work out the mass of metal in a mineral.	
I know that ionic equations can be used to show electrode reactions during electrolysis.	
I know that metals contain positive ions in sea of free-moving electrons.	
I am working at grades B/A/A*	

C6 Grade booster checklist

I know that chemicals have hazard symbols to show that they are hazardous.	
I know that acids react with metals, metal oxides, metal hydroxides and metal carbonates to give a salt and other products.	
I know how to work out the relative formula mass of a compound by adding up the relative atomic masses of the atoms.	
I understand that neutralisation reactions happen when an acid reacts with an alkali.	
I know that some reactions are exothermic and some reactions are endothermic.	
I know that filtration separates solids from liquids or solutions.	
I understand that the rate of reaction is usually measured in amounts per second.	
I know that a catalyst speeds up a reaction without being used up.	
I am working at grades G/F/E	

I know that pure acids can be solids, liquids or gases, and I can name examples of common alkalis.	
I know that reactions of acids can be shown using word and symbol equations with state symbols.	
I know that balanced equations can be used to work out the masses of reactants and products in a reaction.	
I understand that titration results vary, and that the mean is an indicator of the true value of a result.	
I know that acids contain hydrogen ions (H^+) and alkalis contain hydroxide ions (OH^-), and that these react to form water (H_2O) during neutralisation reactions.	
I know that energy-level diagrams can be used to show exothermic and endothermic reactions.	
I know that crystallisation purifies solid crystals and I can list the steps involved.	
I understand that there are several ways of measuring rates of reactions, and how surface area, temperature and catalysts affect the rate of reaction.	
I am working at grades D/C	

I know that the pH scale is used to show the strength of acids and alkalis.	
I know that the formulae of salts can be worked out using the charges on the ions, and that symbol equations balance if the numbers of each type of atom are the same on both sides.	
I can use relative formula masses to work out reacting masses in equations.	
I know that the titration results can be compared if the same concentrations of the same solutions are used.	
I understand how to work out theoretical and percentage yields.	
I know that the gradient on a rate of reaction graph changes as the rate changes.	
I know that the rate of reaction depends on the frequency of collisions of particles.	
I am working at grades B/A/A*	

C7 Grade booster checklist

I know the difference between bulk and fine chemicals.	
I know the main stages in a chemical process.	
I know the differences between exothermic and endothermic reactions and can recognise each type of reaction from an energy level diagram.	
I know why catalysts and enzymes are used in reactions.	
I can use the Periodic Table to find the relative atomic mass of an element.	
I know what hydrocarbons are and what they make when they burn.	
I know the uses of methanol and ethanol and what is made when they burn.	
I understand what happens during fermentation and distillation.	
I know that ethanol can be made from crude oil and biomass.	
I know some examples and properties of carboxylic acids.	
I know why weak acids are less hazardous than strong acids.	
I know that natural esters give fruit and flowers their sweet tastes and smells.	
I know where fats and oils come from and their states at room temperature.	
I know the symbol for a reversible reaction.	
I know what happens in the Haber process and why making ammonia is important.	
I know some advantages and disadvantages of using organic and synthetic fertilisers.	
I know the difference between quantitative and qualitative data.	
I can describe what happens to the mobile phase and stationary phase during chromatography.	
I know how to set up a chromatogram.	
I can describe how gas chromatography works.	
I know how to make a standard solution.	
I know the steps involved in doing a titration.	
I am working at grades G/F/E	

I know that some chemists are involved in research and development in the chemical industry.	
I know the main features of a sustainable chemical process.	
I know how to explain exothermic and endothermic reactions in terms of bonds breaking and forming.	
I know what activation energy means.	
I understand how catalysts and enzymes change the activation energy of a reaction.	
I can use relative atomic masses to work out relative formula masses.	
I understand why alkanes are generally unreactive.	
I know how the properties of alcohols compare with the properties of alkanes.	
I understand why the conditions chosen for fermentation are important.	
I can describe the processes that are used to make ethanol from crude oil and biomass.	
I know the formula for the carboxylic acid group and how carboxylic acids react with metals, alkalis and carbonates.	
I can compare the pH and reactions of strong and weak acids.	
I know the main uses of esters.	
I know the structure of fats and oils.	
I can describe equilibrium reactions in terms of the rate of the forward and backward reactions.	
I understand how changing the conditions affects the rate of reaction and yield of ammonia in the Haber process.	
I can discuss new approaches to making fertiliser that use bacteria and enzymes.	
I can describe how to collect samples so that they represent the bulk of a material under test.	
I understand how substances are distributed between two solvents.	
I know how to compare unknown spots with reference samples on a chromatogram and why locating agents are used.	
I know why different substances have different retention times in a gas chromatogram.	
I can process data by considering outliers, estimates, true values, range, uncertainty and accuracy.	
I know why titration results vary and I can use a given formula to work out concentrations from titration results.	
I am working at grades D/C	

I know some reasons why there are laws governing the chemical industry.	
I know how to use an equation to judge the atom economy of a chemical reaction.	
I understand the advantages and disadvantages of using enzymes in industrial processes.	
I can use bond energy values to calculate the energy given out by an exothermic reaction.	
I can use relative masses and an equation to work out the masses of reactants and products in a reaction.	
I know the difference between saturated and unsaturated compounds.	
I know how alcohols react with sodium and can compare the reaction with that of water and sodium.	
I know why fermentation can only be used to make dilute solutions of ethanol.	
I can discuss the processes that are used to make ethanol by discussing their sustainability.	
I know how the carboxylic acid group forms H^+ ions during reactions.	
I can explain that strong and weak acids produce different concentrations of H^+ ions.	
I know how esters can be made from a carboxylic acid and an alcohol.	
I know the stages in preparing an ester and can describe what happens in each stage.	
I know the differences between saturated and unsaturated molecules.	
I can explain the meaning of the term 'dynamic equilibrium'.	
I understand why the conditions chosen for the Haber process are a compromise.	
I can discuss different processes for making fertiliser based on ideas about sustainability.	
I can describe how to store and prepare samples so that they do not deteriorate before testing.	
I can use ideas about the distribution of substances between the mobile phase and the stationary phase to explain what happens during chromatography.	
I know how to calculate and compare chromatogram R_f values.	
I can interpret print outs from the recorder of a gas chromatogram.	
I can calculate concentrations from volumes and mass of solute.	
I can use information from a titration and an equation to work out concentrations.	
I am working at grades B/A/A*	

Answers

C1 Air quality

Page 89 The changing air around us
1. a Oxygen; nitrogen
   b Mixture
   c Clouds are not made of gases / contain solids and liquids
2. Ratio of 4 nitrogen to 1 oxygen; filling all available space / well spread out
3. Suitable apparatus, e.g. two syringes and a tube containing (excess) copper connected together. Continually pass air over heated copper for a few minutes. Note change in volume and calculate percentage oxygen
4. a Carbon dioxide, water vapour
   b (Earth's) first atmosphere
5. iv before iii; iii before ii; ii before i; i before v
6. Evidence; that is consistent / reliable; little / no evidence to refute the theory; consensus about the explanation of the evidence (Any 2)

Page 90 Humans, air quality and health
1. a Factories; power stations; for transport; in homes (Any 2)
   b Carbon dioxide
   c Air with few pollutants in it / low pollution levels
2. a Goes up / increases each year; more rapidly over last 10 years, consistent / 15 ppm rise between 1960–80; rise is 1990 less than previously (Any 3)
   b i Volcanoes   ii Open fires / vehicle exhausts
   c Sulfur dioxide; nitrogen oxides
   d Asthma; heart disease; lung disease (Any 2)
3. a Stands for parts per million; means 1 gram of pollutants in 1 million grams of air
   b To look for trends; to find areas where pollution is a problem; so people at risk can be warned
4. a Both factors lead to the same outcome
   b Can be caused by other factors/carbon particulates; cannot be certain which one is the cause

Page 91 Burning fuels
1. a Carbon; hydrogen
   b Carbon + oxygen; $\longrightarrow$ carbon dioxide
2. a Oxygen; water; carbon dioxide
   b Oxidation
3. Visual representation to show: $2C_2H_2 + 5O_2 \longrightarrow 2H_2O + 4CO_2$ (1 mark each for: correct representation of molecules; correct equation; balanced)
4. a Molecules   b Rearranged; new
5. a Reactants; products   b The same
6. Atoms cannot be destroyed; they can react / become new substances
7. a Colour change; new substance / gas formed
   b Sulfur is yellow and insoluble; sulfur dioxide colourless and soluble
8. a Contained in plants and animals; so present when fossil fuels forms
   b $S + O_2 \longrightarrow SO_2$ (1 mark for reactants; 1 mark for products)
9. a Acid rain forms when sulfur (and nitrogen oxides) dissolves in water vapour; it damages forests by lowering soil pH; lower pH kills aquatic animals
   b It doesn't affect humans directly

Page 92 Pollution
1. a Power stations; transport (accept examples, e.g. cars)
   b Carbon monoxide   c Sulfur
2. Nitrogen and oxygen (both needed for mark)
3. a Visual representation to show: $2NO + O_2 \longrightarrow 2NO_2$ (1 mark each for correct representation of molecules, correct equation, balanced)
   b Harmful to humans or the environment; any two from: damage buildings; make acid rain; cause breathing / lung problems
4. a Sulfur dioxide   b Carbon dioxide
   c Nitrogen monoxide
5. a Deposited on surfaces   b Acid rain formation
   c Photosynthesis / dissolving in oceans
6. a Reading 4 (8 units)   b 12 to 15
   c $15 + 12 + 14 + 14 + 12 = 67 \div 5 = 13.4$
7. Result might be correct; caused by unusual event (e.g. one very polluting car on a quiet road, gust of wind, etc.)

Page 93 Improving power stations and transport
1. Gas contains less sulfur / makes less sulfur dioxide; easier to remove sulfur from gas before burning; no solid waste products / doesn't make ash; more energy efficient (Any 2)
2. Using flue gas desulfurisation; using alkaline lime slurry; using sea water (Any 2)
3. Switching off devices when not in use / not leaving on stand-by; using newer, more efficient products; accept changing lifestyle (Any 2)
4. Idea of fossil fuels running out / fossil fuels are finite non-renewable resource
5. Wood chips; palm oil; coconut husks; biodiesel; alcohol from sugar (Any 2)
6. a Biofuels are grown as plants; carbon dioxide is trapped in plant as it grows and released when burnt; they are carbon neutral if another plant is grown for every one used
   b Take up land needed to grow food; cannot produce enough to replace fossil fuels
7. Use cars less / walk; use cleaner fuels; remove pollutants from exhaust fumes; use more public transport, make public transport cheaper / more assessable / more frequent (Any 2)
8. a Part of the exhaust system
   b Nitrogen monoxide; carbon dioxide; nitrogen
   c Carbon monoxide gains oxygen and is oxidised to carbon dioxide; nitrogen monoxide loses oxygen and is reduced to nitrogen
9. a Very expensive; not many charging points currently available; takes a long time to charge; limited distance range per charge; dangerous as too quiet for people to hear (Any 3)
   b Most electricity used to charge them is produced at power stations; burning fossil fuels releases carbon dioxide

Page 94 C1 Extended response question
5–6 marks
A detailed description of the patterns shown is given, along with suitable science ideas about causes and why the levels change – an example might be: PM10 levels are lower than NO_x and both vary day to day. A correlation exists: when NO_x levels rise and fall, so do levels of PM10; level rise might be linked to traffic levels / more people travelling weekdays / less at weekends; changes may be due to environmental conditions / wind / rainfall. All information in answer is relevant, clear, organised and presented in a structured and coherent format. Specialist terms are used appropriately. There are few, if any, errors in grammar, punctuation and spelling

3–4 marks
Limited description of the graph and explanation, or a good explanation of one part. For the most part the information is relevant and presented in a structured and coherent format. Specialist terms are used for the most part appropriately. There are occasional errors in grammar, punctuation and spelling

1–2 marks
Little description of patterns, or reasons. Answer may be simplistic. There may be limited use of specialist terms. Errors of grammar, punctuation and spelling prevent communication of the science

0 marks
Insufficient or irrelevant science. Answer not worthy of credit

C2 Material choices

Page 95 Using and choosing materials
1. a Rubber – hard and elastic – car tyres; Plastic – can be moulded – washing-up bowls; Fibres – can be woven – making clothes (Any 2)
   b When talking about materials, a property is something that makes it suitable for the job it does
2. a Melting point
   b Compressive strength
   c Hardness
   d Density
3. a i Climbing ropes need to be dynamic (stretchy) to absorb fall energy and not snap
      ii Climbing ropes need to be dry treated so they do not get heavy when wet and pull the climber off
   b Thicker / 11-mm rope contains more woven fibres; increasing tensile strength
4. a 5.3
   b 5.6 to 5.8
   c To find the true value
   d 5.7
   e An outlier can be discarded if an error occurs in measurement; if one measurement is very different from all the rest

Answers

Page 96 Natural and synthetic materials

1 Good conductor; hard; malleable

2 Clay; glass; cement

3 Polymers

4 a Cotton / paper

  b Silk

  c Limestone / iron ore

5 Synthetic materials are made by a chemical reaction; using raw materials from the Earth's crust

6 Natural materials are in short supply; they can be designed to give particular properties; they can be cheaper; they can be made in the quantities needed (Any 3)

7 Carbon, hydrogen

8 a Fuels

  b Equal

9 a C_4H_{10}

  b C_6H_{14}

10 Due to differences in the marine organisms; that decomposed to make it

11 Diagrams showing correct representation for each molecule; diagram balanced showing two oxygen molecules on the left, and two water molecules on the right

Page 97 Separating and using crude oil

1 a Distillation

  b Evaporates

  c Condenses

  d Fractions

2 The larger the hydrocarbon molecule the higher the boiling point (Accept reverse)

3 Petrol is a smaller molecule; so it has less attractive forces between molecules; so less energy is needed to change it from liquid to gas

4 a Separate paper clips are labelled monomer and joined-up paper clips are labelled polymers

  b i Carbon fibre

    ii Plastic / polythene / polyethene

5

  (Ignore how many – marks given for repeated units; all joined by single bonds)

6 A PET polymer is strong; has low density; does not shatter

7 The properties of polymer chains can be changed by replacing hydrogen atoms; with other groups of atoms

Page 98 Polymers: properties and improvements

1 a i   b i

2 a Low melting point; weak; flexible; soft; insulator (Any 3)

  b Little or no side branches; long chains; strong forces between molecules; high crystallinity (Any 3)

3 Crystallinity holds molecules in regular patterns; the more molecules present the stronger the polymer; and the higher the melting point

4 a Stronger   b More   c Higher

5 Plasticisers are small molecules; that fit between polymer chains; weakening the force between them increases flexibility

6 Thermoplastics have no or few cross-links; so they melt when heated; allowing them to be moulded

7 Natural rubber can be made harder by increasing molecule chain length; and by having more cross-linking

8 a Crystallinity can be increased by drawing the polymer though a small hole to align molecule chains

  b Increased crystallinity makes the plastic brittle

Page 99 Nanotechnology and nanoparticles

1 Microscope

2 10

3 a Salt in sea spray

  b Carbon particulates

  c Designed / made in labs

4 Large molecule

5 One thousand million / one billion

6 Buckyballs are spheres; of 60 carbon atoms. Nanotubes are made from (graphene) sheets; folded into tubes

7 Nanoparticles are effective catalysts because they have a larger surface area; so there are more sites for reactions to occur

8 a Graphene sheets are only one atom thick; but millions of atoms long

  b Volume is the space taken up by a substance. A volume of $1 \, cm^3$ stays the same when cut up. Surface area increases when cut up. The surface area starts at $6 \, cm^2$. When cut in half, it becomes $8 \, cm^2$. When cut into four it becomes $10 \, cm^2$

Page 100 The use and safety of nanoparticles

1 To kill bacteria

2 a 80.5 cm

  b 77 to 79 cm

  c 75 cm after 3 months

  d The range overlaps in each set of results so the true value might be the same; but the means do show it is slightly decreasing. So it is possible it might be getting less bouncy

  e Composites

3 Nanotechnology means the science of making, using and controlling nanoparticles

4 Carbon nanotubes are held together by very strong carbon bonds / bonded like graphite; lightweight as carbon has a low atomic mass; more ductile as while each small tube is stiff, they are not very long, and adjacent ones can flex

5 The silver nanoparticles can be washed into sewage works; and kill the bacteria that clean the water

6 Nanoparticles are small enough to pass through the skin into the body; and possibly lodge in body organs; long-term effects are not known

7 Little actual evidence (so far) of any risk; natural nanoparticles have been around forever without noticed risk; nothing is ever completely safe; people can decide themselves whether to take the risk of using them; ideas that many nanoparticles have useful applications (Any 3)

Page 101 C2 Extended response question

5–6 marks

Answer states at least three properties needed (flexible, lightweight / low density, strong in compression / strong in tension, hard and stiff). Also refers to patterns in the table, e.g. weight and cost links to material used: graphite reduces mass but adds to the cost; nano graphite increases mass as spaces filled in to increase strength and stiffness. All information in answer is relevant, clear, organised and presented in a structured and coherent format. Specialist terms are used appropriately. There are few, if any, errors in grammar, punctuation and spelling

3–4 marks

Answer states three or four properties only, or some properties and a pattern. For the most part the information is relevant and presented in a structured and coherent format. Specialist terms are used for the most part appropriately. There are occasional errors in grammar, punctuation and spelling

1–2 marks

Answer gives one or two properties, and only directly quotes information from the table. Answer may be simplistic. There may be limited use of specialist terms. Errors of grammar, punctuation and spelling prevent communication of the science

0 marks

Insufficient or irrelevant science. Answer not worthy of credit

C3 Chemicals in our lives – risks and benefits

Page 102 Moving continents and useful rocks

1 Geologists

2 Colliding; pulling apart

3 Pangea was formed when many continents collided; forming mountain ranges

4 Over the last 600 million years Britain has moved; weather is linked to (latitude) position / warmer nearer equator

5 Lava erupts; containing magnetic materials; that line up along Earth's magnetic field; their direction shows Earth's magnetic field when formed

6 a The Industrial Revolution started in the north-west of Britain because the industries needed to be where all the raw materials were found

  b Coal; limestone; salt

7 In warm shallow seas shellfish died; compacted and hardened; forming limestone

Answers

8 Salts dissolve in water; enclosed shallow salty sea water evaporates; wind-blown sand becomes mixed with the salt; and buried over time

9 a Ripple marks in the rock; indicate river or wave action

   b Fossils in coal show it is made from plant material; shell fragments in limestone show it is made from sea creatures; shaped grains in rock salt show both water and wind erosion took place *(Any 1)*

Page 103 Salt

1 You can obtain salt by mining it; or by evaporating sea water

2 Salt is put onto icy roads to prevent ice forming; by lowering the freezing point of water; and giving better grip

3 Salt is not obtained from sea water in the UK because it is too expensive to evaporate the water

4 Subsidence could result in large holes underground causing the ground / people's homes to sink into the earth; evaporating salt blown into the environment can damage habitats / pollute water supplies; mining can allow water in mines, which may let salt leach out and contaminate water supplies *(Any 2)*

5 As flavouring; and a preservative

6 As salt concentration increases; the bacteria numbers drop; drop increases at a greater rate after 48%

7 a Salt can cause high blood pressure / heart failure; the risk is estimated by measuring salt intake / by adding totals from labels

   b People can reduce the risk from salt by changing diet to a lower intake; or using 'low salt' alternatives

8 The government regulates food safety through agencies; DH and Defra; the agencies carry out risk assessments; and provide advice to the public

Page 104 Reacting and making alkalis

1 c; d

2 a Potassium sulfate

   b Sodium chloride

3 Stale urine; burnt wood

4 There were few sources; only from burnt wood or seaweed; the demand was greater than the supply

5 a Visual representation to show: $CaCO_3 \longrightarrow CaO + CO_2$ *(1 mark each for correct representation of molecules, correct equation, balanced)*

   b Lime only reacts when acid is present; so any excess just remains in the soil; too much or too little alkali has adverse effects on plant growth; adding the correct amount of alkali is hard to judge; as the amount of acid in a patch of soil is likely to vary *(Any 4)*

6 A mordant sticks dye to fabric

7 a Hydrogen chloride gas; hydrogen sulfide gas

   b Find a use for unwanted products; e.g. as raw materials for a new process

8 a Sodium sulfate + water + carbon dioxide

   b Ammonium nitrate + water

9 Alkalis must be soluble hydroxides; providing hydroxide ions when dissolved

Page 105 Uses of chlorine and its electrolysis

1 a Any number between 90 and 98

   b In Spain the drinking water; is treated with chlorine

2 They may have been infected on holiday; or by drinking untreated water

3 a Chlorine reacts with organic materials in water; forming toxic or carcinogenic compounds / disinfectant by-products (DBPs)

   b Amounts of DBPs are very small; so the risks are small; compared to the risk of cholera or typhoid; in this case the benefits outweigh calculated risks;

4 a Electrolysis

   b Hydrogen; chlorine; sodium hydroxide

   c All the products are useful (so there is no waste)

   d Making these products requires large amounts of electricity or energy; to melt the electrolyte; and to separate using electrolysis

   e Chlorine is used for PVC / medicines / crop protection; hydrogen is used for margarine / rocket fuel / fuel cells; sodium hydroxide is used for paper making / in domestic cleaners / for refining aluminium

5 Environmental impacts plus potential solutions:

   • Chlorine is linked to ozone depletion; it was used in fridges and aerosols; but has now been banned. Collection points have been set up by local authorities so fridges can be disposed of safely

   • Chlorine is linked to dioxins in paper bleaching; which are linked to cancer. Manufacturers try to find alternative for bleaching; or ways to find to prevent dioxins escaping

   • Mercury is linked to toxic poisoning; which can build up over time in animals higher up the food chain. Laws are in place in some countries to ban the mercury method of electrolysis. You could also use an alternative, more efficient membrane cell method *(Any 1)*

Page 106 Industrial chemicals and LCA

1 a Lead could spread to the environment from dumping; escapes during manufacture or mining; is released into the air if batteries are burnt *(Any 2)*

   b Recycling lead batteries reduces the need for processing or mining new lead

   c A cumulative poison is when something builds up; slowly over time; in body tissues

2 People *perceive* a greater risk with a less familiar named chemical; long-term studies to measure actual risk are not possible; the *actual* risk may be very small or non-existent; the EU believe that if there might be a risk it should be avoided

3 a Carbon

   b Chlorine

   c Hydrogen

4 a It will not break down naturally

   b Plasticisers

5 Idea that they leach out easier; unknown health risk

6 An LCA measures energy use and environmental impact; from when something is made; until its disposal

7 The LCA is better for the table: though more energy is used to make a table; it has a longer life and may never be disposed of; the table has less environmental impact; than burning wood which releases carbon dioxide

8 A full LCA is not always possible because of insufficient data; it is too hard to measure some aspects; e.g. a product life / the disposal method is often unknown

Page 107 C3 Extended response question

5–6 marks

LCA defined as the energy needed to make it, use it, dispose of it. Answer contrasts the two charts effectively using examples, e.g.:

• large amount of energy needed for extracting the oil, refining it, cracking into monomers and making the polymer;

• energy needed to collect plastics disposed off, and to clean and shred, but less than making from new;

• LCA during the life of an individual item the same, but recycling has better LCA as it extends the life and reduces the need to burn or go to landfill.

All information in answer is relevant, clear, organised and presented in a structured and coherent format. Specialist terms are used appropriately. There are few, if any, errors in grammar, punctuation and spelling

3–4 marks

LCA defined with some contrast of the LCA and some explanation. For the most part the information is relevant and presented in a structured and coherent format. Specialist terms are used for the most part appropriately. There are occasional errors in grammar, punctuation and spelling

1–2 marks

'Cradle to grave' stated with some reference to differences, although not explained. Answer may be simplistic. There may be limited use of specialist terms. Errors of grammar, punctuation and spelling prevent communication of the science

0 marks

Insufficient or irrelevant science. Answer not worthy of credit

Answers

C4 Chemical patterns

Page 109 Atoms, elements and the Periodic Table

1 Relative atomic masses / properties / specific examples of properties, e.g. reactivity / melting points / density (*Any 2 = 1 mark each*)

2 a $(7 + 39) \div 2 = 22.5$

  b The RAM of sodium is roughly the mean of the other two

3 a They have similar properties

  b It left gaps for undiscovered elements; made predictions about the properties of elements

4 Looking at light from the Sun; splitting light using a prism; looking at the patterns of lines; idea that the pattern is unique to an element / element can be identified from the pattern (*Any 3*)

5 Protons / neutrons in either order; electrons; shells

6

Proton number	Relative atomic mass	No. of protons	No. of neutrons	No. of electrons
9	19	9	10	9
13	27	13	14	13
3	7	3	4	3

(1 mark for each correct row)

Page 110 Electrons and the Periodic Table

1 a 2 dots in first shell, 8 in second, one in third; electron arrangement: 2.8.1

  b Both have 1 electron in their outer shell; potassium has more electrons than sodium / 19 instead of 11; potassium has more electron shells than sodium / 4 shells instead of 3

2 a

Name of element	Symbol	Proton number
Neon	Ne	10
Fluorine	F	9
Lead	Pb	82

(1 mark for each correct row)

  b Aluminium

3 c; d

4 The number of electrons in the outer shell is the same as the group number; elements with 3 or fewer electrons in the outer shell are usually metals; elements with 5 or more electrons in the outer shell are non-metals

Page 111 Reactions of Group 1

1 a Decreases down the group

  b 850–1200°C (must have units); boiling point increases down the group

2 a a to iii; b to i; c to ii

  b Both fizz / produce hydrogen; both turn pH indicator blue / make an alkali / make a hydroxide; potassium reaction is faster / produces a flame; produces potassium hydroxide rather than sodium hydroxide

3 a $2Li + Cl_2 \longrightarrow 2LiCl$ (*$Cl_2 = 1$ mark, balancing = 1 mark*)

  b Lithium has 2 electron shells / electron arrangement 2.1. Sodium has 3 electron shells / electron arrangement 2.8.1. Potassium has 4 electron shells / electron arrangement 2.8.8.1. Reactivity increases down the group; number of electron shells also increase down the group (*Any 4 points*)

Page 112 Group 7 – The halogens

1 Solid – iodine – dark grey; liquid – bromine – red-brown; gas – chlorine – pale green (*States all correct = 1 mark, only two colours correct = 1 mark, all three colours correct = 2 marks*)

2 Cl_2; contains two atoms in each molecule

3 a Gases at the top, then a liquid, then solids at the bottom

  b Boiling points increase; melting points increase; density increases; colour becomes darker; reactivity decreases (*2 marks for any one answer*)

4 a Iron chloride

  b Rose should do the experiment in a fume cupboard; chlorine is a toxic gas

  c Rate of the reaction is slower; bromine is less reactive

5 a $Cl_2 + 2KAt \longrightarrow 2KCl + At_2$ (*$At_2 = 1$ mark, balancing = 1 mark*)

  b Chlorine has fewer electron shells; Group 7 / non-metals are more reactive when they have fewer electron shells

Page 113 Ionic compounds

1 b; e

2 a C

  b A

---

3 a Below melting point it does not conduct; above melting point its conductivity increases / it does conduct

  b Ions cannot move in the solid; when the solid melts the ions can move

4 a Both have the same number or 11 protons; both have same number or 12 neutrons; ion has fewer electrons / loses an electron / has 10 not 11 electrons

  b A sodium atom loses an electron; a chlorine atom gains an electron

5 Sodium ions have a charge of +1; chloride ions have a charge of –1; idea that charges balance one to one in sodium chloride; the charge on a calcium ion must be +2; need two chloride ions to balance the charge in calcium chloride (*Any 4 points*)

Page 114 C4 Extended response question

5–6 marks
Answer makes clear comparisons and discusses both lithium and chlorine and gives information about electron arrangement, metal/non-metal character and reactivity. Possible marking points:
- lithium has 1 electron in the outer shell
- fluorine has 7 electrons in the outer shell
- lithium is a metal
- chlorine is a non-metal
- lithium is less reactive than other elements in Group 1
- fluorine is more reactive than other elements in Group 7.

All information in answer is relevant, clear, organised and presented in a structured and coherent format. Specialist terms are used appropriately. There are few, if any, errors in grammar, punctuation and spelling

3–4 marks
Answer discusses both lithium and chlorine and discusses at least two from: electron arrangement, metal/non-metal character and reactivity. For the most part the information is relevant and presented in a structured and coherent format. Specialist terms are used for the most part appropriately. There are occasional errors in grammar, punctuation and spelling

1–2 marks
Answer gives some properties of lithium and/or chlorine but comparison is not clear. Answer may be simplistic. There may be limited use of specialist terms. Errors of grammar, punctuation and spelling prevent communication of the science

0 marks
Insufficient or irrelevant science. Answer not worthy of credit

C5 Chemicals of the natural environment

Page 115 Molecules in the air

1 O_2 – oxygen – 21%; Ar – argon – about 1%; CO_2 – carbon dioxide – 0.04%; N_2 – nitrogen – 78%

2 b; c

3 a Oxygen

  b The forces between molecules are very weak; so a small amount of energy can overcome the forces

  c Nitrogen has the lowest boiling point; but it does not have the lowest melting point. Oxygen has the lowest melting point; but oxygen does not have the lowest boiling point. The pattern is not true for any of the three gases

Page 116 Ionic compounds: crystals and tests

1 a Positive and negative ions; are very strongly attracted together; and need large amounts of energy to overcome the forces between them

  b Ions cannot move

  c Ions move about; become further apart / are not arranged regularly/ spread through the water

2 Formula for calcium chloride: $CaCl_2$. Potassium sulfate: positive ion – K^+ and negative ion – SO_4^{2-}; formula – K_2SO_4

3 a Calcium carbonate

  b Similarity – still get a white precipitate; difference – this time it dissolves in excess

  c Insoluble

  d $Ca_2^+(aq) + 2OH^-(aq) \longrightarrow Ca(OH)_2(s)$ (*correct equation = 1 mark, state symbols = 1 mark*)

Page 117 Giant molecules and metals

1 a 12%

  b Silicon and oxygen (*both = 1 mark*)

  c Reduction

2 Only very small amounts of copper are found in the rock; so to make a small amount of copper, large amounts of waste rock are produced

Answers

3 a Simple molecular structure contains small molecules; idea that only a few atoms are joined in each molecule; idea that many atoms are joined in a giant covalent structure; idea that the atoms are joined in 3 dimensions *(Any 3 points)*

   b i False ii True iii True iv False *(All 4 correct = 2 marks; 2 or 3 correct = 1 mark)*

Page 118 Equations, masses and electrolysis

1

	iron oxide	carbon monoxide	iron	carbon dioxide
reactant (✓)	✓	✓		
product (✓)			✓	✓

2 Copper; carbon dioxide and CO_2

3 a RAM of Mg: 58.5; RFM of NaCl: 24; RFM of $CaSO_4$: 136

   b 95 g *(must have g)*

4 a b

   b 54% *(Uses correct atomic masses 27 and 16 = 1 mark, correct answer = 1 mark)*

   c i Oxygen ions are negatively charged; negatively charged ions are attracted to the positive electrode

      ii $Al^{3+} + 3e^- \longrightarrow Al$ *(3e = 1 mark, rest correct = 1 mark)*

Page 119 Metals and the environment

1 a i Excellent electrical conductivity

      ii Expensive; corrodes quickly

   b Excellent conductivity; very light in weight

2 a Reducing the energy needed for lighting is a benefit; it is impossible to stop mining mercury altogether because we need mercury to make light bulbs; people buy low-energy light bulbs because they think it helps the environment; carbon dioxide causes climate change; if the mines were closed people would lose their jobs; the costs of the mines include the toxic water; the costs include the toxic gases that the production produces; it might be possible to reduce the environmental impact of the mercury mine *(Any 4)*

   b The sea of electrons in the metal; can move

Page 120 C5 Extended response question

5–6 marks
Discusses and compares similarities and differences for both diamond and silicon dioxide and gives details of the structure and state (solids) of both. Possible points:
- both solids / high melting points
- both hard
- both regular arrangement or lattice
- both covalently bonded; both giant / 3-D structures
- diamond is an element / only contains carbon atoms
- silicon dioxide contains both silicon and oxygen atoms / it is a compound.

All information in answer is relevant, clear, organised and presented in a structured and coherent format. Specialist terms are used appropriately. There are few, if any, errors in grammar, punctuation and spelling

3–4 marks
Covers some points about the structure of both diamond and silicon dioxide. Gives at least one similarity and one difference between the two. For the most part the information is relevant and presented in a structured and coherent format. Specialist terms are used for the most part appropriately. There are occasional errors in grammar, punctuation and spelling

1–2 marks
Gives some information about either silicon dioxide or diamond but does not make a comparison clear. Answer may be simplistic. There may be limited use of specialist terms. Errors of grammar, punctuation and spelling prevent communication of the science

0 marks
Insufficient or irrelevant science. Answer not worthy of credit

C6 Chemical synthesis

Page 121 Making chemicals, acids and alkalis

1 a Flammable hazard symbol

   b No naked flames; keep the top on the can; be careful not to spill any *(Any 2)*

2 a Sulfuric acid (l); citric acid (s); hydrochloric acid (g)

   b Sulfuric acid

   c Hydrogen

3 a Copper chloride, carbon dioxide and water; 2HCl; CO_2 and H_2O

   b pH starts off low / at pH 1; pH rises when copper carbonate is added; because the acid is neutralised

   c $Cu(NO_3)_2$

Page 122 Reacting amounts and titrations

1 MgO; $(23 \times 2) + 16 = 62$; sodium carbonate; $(2 \times 23) + 12 + (3 \times 16) = 106$

2 a Same number of atoms idea; on both sides of the equation

   b Relative atomic masses: S = 32 and O = 16; relative formula mass $ZnSO_4$ = 161; mass of zinc sulfate = $161 \times 2 = 322$ g

3 a Add sodium hydroxide until indicator changes colour; with shaking or stirring; add drop by drop near the end-point; do repeats; check that repeats are close together; take an average / mean of the repeats *(Any 4)*

   b a = True; b = False; c = True; d = False *(All correct = 2 marks; 3 correct = 1 mark)*

Page 123 Explaining neutralisation and energy changes

1 Hydrochloric acid, potassium hydroxide, potassium chloride; nitric acid, sodium hydroxide, sodium nitrate; sulfuric acid, calcium hydroxide, calcium sulfate

2 Water

3 a Hydrogen, H^+ b Hydroxide, OH^-
   c $H^+ + OH^- \longrightarrow H_2O$

4 Increases; given out; neutralisation

5 Energy is given out; the reactants have more energy than the products

6 Idea of very large temperature rise; causing explosion / fire / hazard

Page 124 Separating and purifying

1 a Calcium nitrate

   b Filter; the solid stays on the filter paper / does not go through

   c a = True; b = False; c = False; d = True *(All correct = 2 marks; 3 correct = 1 mark)*

2 a More soluble / more dissolves / dissolves faster *(Any 2)*

   b Step 2 takes out insoluble impurities / impurities that do not dissolve; Step 4 filters off the crystals

   c Drying; in an oven or dessicator

3 a RFM of $ZnCl_2$ = 65 +(2×35.5) = 136. Theoretical yield = 13.6 g

   b 10.2 / 13.6 × 100; = 75%

   c It will be too high; due to extra mass of water

Page 125 Rates of reaction

1 a Volume of acid; mass of calcium carbonate; temperature *(Any 2)*

   b As the concentration increases; the rate of reaction increases

   c Idea that is it too short a time; it should be about 200 s / ten times slower than Experiment 2 / it does not fit the pattern

2 a Every 30 s   b 5 minutes

   c The gradient becomes less steep; because the reaction slows; it becomes level when the reaction stops; because the reactants are being used up; collisions become less frequent; less frequent collisions lead to a slower reaction *(Any 4)*

Page 126 C6 Extended response question

5–6 marks
Links shape of graph to rate of reaction and gives explanation for the changes, e.g. a gas is made and the acid is used up. Points to include:
- mass decreases during the reaction; due to loss of a gas
- (describes shape of graph) graph changes fast at first and then levels out
- (links shape to rate of reaction) reaction is fastest at first and then slows
- reactions stops when line goes flat
- reaction slows when acid concentration falls
- reaction stops when acid is used up.

All information in answer is relevant, clear, organised and presented in a structured and coherent format. Specialist terms are used appropriately. There are few, if any, errors in grammar, punctuation and spelling

3–4 marks
Links shape of graph to rate of reaction and discusses reaction slowing and stopping. For the most part the information is relevant and presented in a structured and coherent format. Specialist terms are used for the most part appropriately. There are occasional errors in grammar, punctuation and spelling

Answers

1–2 marks
Describes how the shape of the graph changes over time.
Answer may be simplistic. There may be limited use of specialist
terms. Errors of grammar, punctuation and spelling prevent
communication of the science

0 marks
Insufficient or irrelevant science. Answer not worthy of credit

C7 Further chemistry

Page 127 The chemical industry

1 a F, T, F, F (*2 marks for all correct, 1 mark for 2 or 3 correct*)

 b Large, small, small, large (*2 marks for all correct, 1 mark for 2 or 3 correct*)

2 Box 2; box 3

3 a To protect people / workers; to protect the environment; give an example of a specific hazard, e.g. chemicals may be flammable, toxic, harmful (*Any 2*)

 b Identify nature of hazard / if it is flammable or toxic or harmful; what to do if there is a fire; what to do if there is a spillage; contact details for more information; labels have information for the fire service / give an example of information important to the fire service (*Any 2*)

Page 128 Green chemistry

1 a A, E, B, D, C (*2 marks for all correct, 1 mark for A first and C last*)

 b E

2 a Corn is a renewable resource / crude oil is not renewable; waste rots away / does not cause a long-term litter problem

 b If enough corn can be grown to supply demand for new process; energy inputs / outputs; if energy used in process is from renewable sources; what by-products or waste is produced; health / safety / social economic issues relating to people; effect on the environment of the process (*Any 3*)

3 a F, T, F, F, T (*2 marks for all correct, 1 mark for 4/5 correct*)

 b More sustainable; because there is less waste

Page 129 Energy changes

1 a Reaction A is exothermic; because reactants are higher than products / energy given out during reaction; reaction B is endothermic; because reactants are lower than products / energy taken in during reaction

 b Reaction B (*no marks*); because a fall in temperature to the surroundings show energy is taken in

2

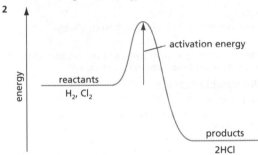

 a Line starting level with reactants, with arrow meeting highest part of hump, labelled 'activation energy'

 b Flash or spark provides activation energy; to break bonds to start reaction; reaction is exothermic; reaction provides energy for bonds to continue to break to continue the reaction

 c It is lower; energy is not needed / enough energy available at room temperature to start reaction

Page 130 Catalysts and enzymes

1 a To make the reaction faster; so that a lower temperature can be used / to make more margarine in a fixed time

 b They do not get used up in the reaction

 c Nickel catalyst is not affected in the reaction / still there at the end; nickel catalyst is a solid / hydrogenated vegetable oil forms as a gas

2

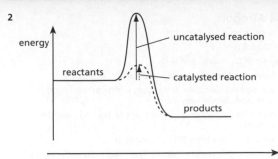

 a i Two 'humps' shown between reactants and products; one 'hump' clearly higher than the other

 ii Arrow from level of reactants to top of higher hump labelled 'activation energy for uncatalysed reaction'; arrow from level of reactants to top of lower hump labelled 'activation energy for catalysed reaction'

 b Lower activation energy with a catalyst; more particles have enough energy to react

 c Reaction happens at lower temperature / save energy / enzymes are biodegradable / can make reactions with high atom economy work; only work in narrow temperature / pH range / may denature if conditions are not suitable

Page 131 Energy calculations

1 a Activation energy

 b To start the reaction; by breaking bonds

 c Different reactants have different bonds; different bonds have different bond energies

 d B; because energy is given out / shows exothermic change

 e i Exothermic change

 ii Energy taken in when bonds break; energy given out when bonds form; energy given out is more than energy taken in

2 a Energy given out when bonds form = $4 \times 805 = 3220$ kJ/mol; energy taken in when bonds break = $(2 \times 1079) + 498 = 2656$ kJ/mol; energy change of reaction = $3220 - 2656 = 564$ kJ

 b Energy taken in when bonds break is lower than energy given out when bonds form; energy given out overall

Page 132 Reacting masses

1 a

Symbol	Atomic number	Relative atomic mass
He	2	4
C	6	12
F	9	19
Mg	12	24

(*1 mark for each correct row*)

 b Relative or compared to carbon; one carbon atom is given a mass of 12

 c Most elements in the table have a relative atomic mass double the atomic number; these are He, C and Mg; but F does not obey the rule / relative atomic mass is not double the atomic number

 d Needs to collect atomic numbers and relative atomic masses; for other elements / all elements on the Periodic Table

2 a $HCl = 36.5$; $MgCl_2 = 95$; $H_2 = 2$

 b 24 g Mg make 95 g $MgCl_2$ / allow ecf from table; 6 g makes 23.75 g

Page 133 Alkanes

1 a i Contains hydrogen and carbon; ONLY

 ii C_2H_6

 b Oxygen; carbon dioxide; water

 c Ethane contains all single bonds; which are difficult to break; therefore reactions have high activation energies

2 a i C_3H_8

 ii Both hydrogen and carbon / are hydrocarbons / have 6 hydrogen atoms; different numbers of carbon atoms / ethane has fewer carbon atoms / ethane has 2 carbon atoms, propene has 3 / ethane has all single bonds and propane has a double bond / ethane is saturated and propene is unsaturated

 b Saturated contains only C—C and C—H single bonds; unsaturated contains double carbon-carbon bonds

Answers

Page 134 Alcohols

1 a i Fuels; solvents

   ii Methanol is toxic

   iii Oxygen; CO_2; water and H_2O

2 a Both have a chain of hydrogen and carbon atoms

   b No because ethanol contains oxygen; hydrocarbons contain hydrogen and carbon only

   c i (Ethanol has) a higher melting point or boiling point / is less flammable

   ii Ethanol contains the –OH functional group

3 a Fizzes in water; fizzes in ethanol; fizzes more slowly in ethanol than in water; no reaction with hexane

   b hydrogen; sodium ethoxide

Page 135 Fermentation and distillation

1 a Yeast

   b Sugar

   c Ethanol

   d Carbon dioxide

   e Mixture is heated until it boils; vapour is cooled / condensed; liquid is more concentrated / contains more alcohol / ethanol

2 a Sugar is needed to give food to the yeast; yeast provides enzymes that make fermentation happen; air must be kept out because yeast respires anaerobically; enzymes have optimum temperature and do not work if too hot or cold

   b Carbon dioxide

3 Ethanol is made by fermentation; ethanol is toxic; yeast dies at high ethanol concentrations

Page 136 Alternatives to fermentation

1 Waste straw from crops; crude oil

2 a ethane ⟶ ethene + hydrogen; ethene + water / steam ⟶ ethanol

   b Reaction uses a high temperature

   c i Only hydrogen is made as a waste product; most of the atoms from ethane end up in the ethanol product; refers to percentage by mass of atoms

   ii Hydrogen is used as a fuel / making margarine / any correct use of hydrogen

3 a Both use plant material; both raw materials are renewable; both use microorganisms; both involve microorganisms using plant material as food / for respiration; both involve enzymes; both work at just above room temperature (Any 2)

   Plus any 2 points from: fermentation uses a natural micro-organism; *E-coli* is genetically engineered; fermentation uses sugar; bacteria work on waste plant material / biomass

   b Fermentation uses only a small amount of the plant material / makes a mixture that needs further separation / makes only a dilute solution of ethanol / produces a lot of waste / makes carbon dioxide as a waste product; using biomass and bacteria is not developed for large scale use / uses acids and solvents / needs distillation to produce ethanol

Page 137 Carboxylic acids

1 a Methanoic acid; CH_3COOH; vinegar

   b Sharp taste; smell

2 a Ring COOH

   b magnesium ethanoate; hydrogen

   c Sodium hydroxide

   d Sodium methanoate; CH_3COONa; $CH_3COO^- Na^+$; (2)CH_3COO^- Mg^{2+}

3 a Both produce H^+ ions

   b The substances are dissolved in water

Page 138 Carboxylic acids

1 a Methanoic acid; ethanoic acid

   b Safe at low concentrations / only harmful at high concentrations; acids are corrosive

   c Sulfuric acid is a strong acid; carboxylic acids are weak acids; strong acids are more corrosive / more hazardous

2 a Use pH paper / pH probe / universal indicator; ethanoic acid has a higher pH than hydrochloric acid

   b Similarities (can score up to 2 marks): both make a gas / both fizz; both make carbon dioxide; both make a salt

   Differences: ethanoic acid reacts more slowly than hydrochloric acid / fizzes more slowly; salt is different for each

3 a $HCOOH$; ⟶ $H^+ + HCOO^-$

   b Both produce H^+ ions; hydrochloric acid ionises completely; only some methanoic acid molecules ionise

Page 139 Esters

1 a Flowers are not the only source; also found in fruit; give flowers their smell (not colours)

   b i Act as solvents

   ii Low boiling points / evaporate easily; contain a hydrocarbon chain / contain carbon, hydrogen and oxygen atoms / are organic compounds

   c Act as plasticisers; make plastic more flexible

   d They are cheaper to make / natural esters are more difficult / expensive to source / natural esters not available in large quantities

2 a

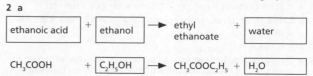

   b Heat; catalyst used; catalyst is concentrated sulfuric / phosphoric acid

Page 140 Making esters

1 a i C, B, E, A (2 marks for all correct, 1 mark for C first and A last)

   ii C

   iii ethanoic acid; H_2SO_4; acts as a catalyst

   b Gases allowed out of top of funnel; liquids separate and (lower) layer can be run off

   c Anhydrous; calcium chloride

2 Both use a condenser / both involve heating; reflux uses a vertical condenser; distillation uses a condenser sloping downwards reflux is used to stop substances evaporating / returns substances to flask; distillation is used to purify the product / separate the ester from the reaction mixture

Page 141 Fats and oils

1 T, F, F, T (2 marks for all correct, 1 mark for 2 or 3 correct)

2 a Ring any O-C=O group (no other atoms inside the ring)

   b Contain three ester groups

   c i H_2O

   ii

   ```
       H
       |
   H−C−OH
       |
   H−C−OH
       |
   H−C−OH
       |
       H
   ```

   Three –OH groups correct; rest of molecule correct; glycerol

   d Box 3

   e Oil A contains only single bonds between carbon atoms; olive oil contains one or more double carbon-carbon bonds

Page 142 Reversible actions

1 a Reaction is reversible

   b The reaction can go backwards as well as forwards; carbon dioxide dissolved in water / CO_2(aq) can reform carbon dioxide gas / CO_2(g)

   c Closed container means that reaction will stay at equilibrium; amount of carbon dioxide dissolved in water / CO_2(aq) will stay the same

2 a So that the reaction reaches equilibrium; because no reactants or products can escape

   b Rate of forward reaction; is the same as rate of reverse reaction; amount of reactants and products stay constant

   c i Reaction reaches equilibrium

   ii Both decrease; they are used up in the reaction; stop decreasing / stay constant at equilibrium

Page 143 The Haber process

1 a Box 3

   b Plants need nitrogen for growth; not enough available in soil / crops use nitrogen up

2 a Decreases

   b Increases

   c i Low temperature; high pressure

   ii Too high pressures are expensive to maintain / too many leaks; very low temperatures mean that the reaction is too slow

   d Increases surface area of catalyst; to make reaction even faster / more gas molecules in contact with surface

Answers

Page 144 Alternatives to Haber

1 a

	True only for synthetic fertilisers	True only for organic fertilisers	True for both
Contain nitrogen compounds			✓
Made from feedstocks	✓		
Made from waste material		✓	

  b i Growth of algae; bacteria multiply; bacteria use up all of the oxygen; fish / other living things die *(Any 2)*

  ii No, because synthetic fertilisers get washed into rivers; but organic fertilisers also get washed into rivers

2 a Haber: uses natural gas; energy for process is generated from fossil fuels; needs high temperature; needs high pressure

  Enzyme: air and water are both renewable; less energy needed because uses room temperature *(Any 5)*

  b Not developed for large scale use / only experimental

3 Not enough fertiliser available; to meet demand / cannot grow enough vegetables

Page 145 Analysis

1 a

	Quantitative data	Qualitative data
What metal ions the water contains		✓
Concentration of dissolved oxygen	✓	
Water temperature	✓	
Types of algae in the water		✓
Rate of flow of the river	✓	

  (2 marks for all correct, 1 mark for 4/5 correct)

  b i Take multiple samples; when factory is operating; when factory is not operating; from different positions in the river *(Any 2)*

  ii Wear gloves; use clean equipment to collect samples; use clean containers for storage; use containers that are made from materials that will not contaminate samples / do not contain plasticisers *(Any 2)*

  c Repeating tests identifies outliers / can take mean of the results; standard procedures mean results of tests can be compared / means the test is the same every time

  d Keep dark; in the fridge / in cold conditions / do not freeze

Page 146 Principles of chromatography

1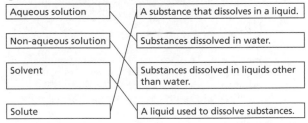

Aqueous solution	— A substance that dissolves in a liquid.
Non-aqueous solution	— Substances dissolved in water.
Solvent	— Substances dissolved in liquids other than water.
Solute	— A liquid used to dissolve substances.

2 a Iodine is more soluble in hexane than in water; distribution is that more iodine is in the hexane layer than in the water layer; evidence: iodine dissolved in hexane has a deeper colour linked to iodine more soluble in hexane; iodine dissolved in water has a pale colour linked to iodine less soluble in water; solid iodine left when shaken with water / no iodine left when shaken with hexane *(Any 4)*

  b Hexane is less dense; violet layer is on the top

3 Inks / substances dissolve in mobile phase; inks / substances are carried along / moved by mobile phase; spread out due to different solubilities; inks / substances have moved further when water is used; inks / substances are more soluble in water than in ethanol *(Any 4)*

Page 147 Paper and thin-layer chromatography

1 B, D, A, E *(1 mark for all correct, 1 mark if step C is left out)*

2 a X contains safe dyes A and B and so is safe; Y contains safe dye C; Y also contains an unknown dye; you cannot tell if the unknown dye in Y is safe; neither food colouring contains the banned dye *(Any 4)*

  b Correct answer is 0.6; if correct answer not scored, marks allowed for:

  distance travelled by spot = 3 / distance travelled by solvent is 5;

  R_f value = $\dfrac{\text{distance moved by sample}}{\text{distance moved by solvent}}$

  c Colourless samples

Page 148 Gas chromatography

1 a

Part	Purpose
Carrier gas	Mobile phase/carries sample through column.
Solid in the column	Acts as stationary phase.
Oven	Controls temperature / heats the column.
Detector	Detects the substances as they leave the column.
Recorder	Gives a graph showing a peak for each substance.

  b It is unreactive / inert; doesn't react with the substances

2 a i Oil D

  ii Weaker attractions than other oils / they are weak; because the oil travels through the column the fastest

  b Oil C; highest peak

  c Do chromatograms of samples of margarine from each company; do chromatogram of unknown margarine; compare the print outs; look for heights of peaks / position of peaks

  d Measures mass of substances; identifies substances

Page 149 Quantitative analysis

1 a Volumetric flask

  b Dissolve the solid in a beaker in (distilled) water; pour into the volumetric flask; wash the beaker and add the water to the flask; make the solution up to the line; make sure the bottom of the meniscus is on the line; invert the flask *(Any 4)*

2 a i Sample 4 / 4.6

  ii 3.5 g *(2 marks for correct answer, allow 1 mark if 4.6 has been included in calculation to give 3.68 g)*

  b i 3.3−3.7 g

  ii Both sets agree with each other; range is narrow

3

Solution	Mass of solute in g	Volume of solution	Concentration in g/dm³
A	20	0.5 dm³	40
B	10	200 cm³	50
C	3	250 cm³	12

Page 150 Acid-base titration

1 Put an acid in a burette; use a pipette to measure the NaOH; add an indicator to the NaOH; add the acid to the NaOH; stop adding acid when indicator changes colour *(Maximum of 4)*

  Details: use a white tile; use the meniscus when measuring; do repeats / do a rough / take averages of results; add dropwise near the end *(Maximum of 2)*

2 Rough is not done carefully; not reading burette or pipette properly; adding too much acid / going past end point; difficult to see colour change; losing drops of solutions / splashes *(Any 2)*

3 a Fizzing happens because a gas is made; which is carbon dioxide

  b Na_2CO_3 106; HCl 36.5

  c Mass of HCl used =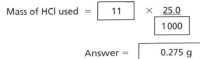

  Answer = 0.275 g

  Correct answer: 0.275 g *(2 marks for correct answer, 1 mark for 11 or 1000 correctly entered in working)*

  d 106 g Na_2CO_3 reacts with 73 g HCl *(if 36.5 g HCl is given, allow error carried forward in rest of calculation)*; 106 × 0.275 *(allow ecf if incorrect values given in b and c for 73; 0.399 g / 0.4 g)*

Answers

Page 151 Extended response question

Relevant points include:

Strength of acids
- Sulfuric acid and hydrochloric acids are strong acids.
- Ethanoic acid and citric acid are weak acids.
- Citric acid is a stronger acid than ethanoic.

Evidence
- Stronger acids react faster with sodium carbonate / fizz more quickly.
- Weak acids fizz slowly / idea that rate of fizzing indicates how strong acid is.
- Stronger acids react faster with magnesium / magnesium disappears more quickly.
- Weak acids react with magnesium slowly / idea that rate of disappearance of magnesium indicates how strong acid is.
- Strong acids have low pH / weak acids have higher pH.
- Identifies difference in pH between ethanoic acid and citric acid as important

5–6 marks

Identifies each acid as strong or weak and discusses the reactions with sodium carbonate, magnesium and refers to pH as evidence. Specialist terms are used appropriately and all information in answer is relevant, clear, organised and presented in a structured and coherent format. There are few, if any, errors in grammar, punctuation and spelling.

3–4 marks

Identifies each acid as strong or weak and links some of the information in the table. For the most part the information is relevant and presented in a structured and coherent format. Specialist terms are used for the most past appropriately. There are occasional errors in grammar, punctuation and spelling.

1–2 marks

Recognises the strength of the acids as strong or weak but may not relate this to the evidence in the table. Answer may be simplistic. There may be limited use of specialist terms. Errors of grammar, punctuation and spelling prevent communication of science.

0 marks

Insufficient or irrelevant science. Answer not worthy of credit.